DISNEY DEMYSTIFIED

Volume Two

The Stories and Secrets
Behind Disney's Favorite
Theme Park Attractions

David Mumpower

Theme Park Press
The Happiest Books on Earth
www.ThemeParkPress.com

© 2017 David Mumpower

No part of this publication may be reproduced, distributed, or transmitted in any form or by any means, including photocopying, recording, or other electronic or mechanical methods, without the prior written permission of the publisher, except for brief quotations embodied in critical reviews and certain other non-commercial uses permitted by copyright law.

Although every precaution has been taken to verify the accuracy of the information contained herein, no responsibility is assumed for any errors or omissions, and no liability is assumed for damages that may result from the use of this information.

Theme Park Press is not associated with the Walt Disney Company.

The views expressed in this book are those of the author and do not necessarily reflect the views of Theme Park Press.

Theme Park Press publishes its books in a variety of print and electronic formats. Some content that appears in one format may not appear in another.

Editor: Bob McLain
Layout: Artisanal Text

ISBN 978-1-68390-088-7
Printed in the United States of America

Theme Park Press | www.ThemeParkPress.com
Address queries to bob@themeparkpress.com

Contents

Introduction v

ONE
Disney Landmarks 1

TWO
World Showcase 17

THREE
Disney Spies 37

FOUR
The Living Seas 53

FIVE
Hollywood Studios 73

SIX
The Osborne Family
Spectacle of Dancing Lights 95

SEVEN
Animal Kingdom 111

EIGHT
Beastly Kingdom 129

NINE
Pleasure Island 149

TEN
How "Frozen" Took Over the World   167

ELEVEN
The Fifth Disney Park   173

TWELVE
Disney Myths and Legends   189

Acknowledgments   207
About the Author   209
More Books from Theme Park Press   211

Introduction

As I discussed in detail in *Disney Demystified: Volume One*, Imagineers faced a seemingly impossible situation in the wake of Walt Disney's death. The man who built the Disney dynasty had spent the last years of his life hoarding Florida swampland in anticipation of building the Experimental Prototype City of Tomorrow. He imagined a utopia based on capitalism and innovation.

What Walt couldn't have possibly anticipated when he announced the project in November of 1965 was that he would die only 13 months later. He was the man with the dream and the ambition about the Florida Project. His death left a void that was difficult to fill.

Roy O. Disney immediately demonstrated loyalty to his late brother by putting off retirement until the completion of what everyone now called Disney World, later expanded into Walt Disney World. This new Florida community was intended as the loving tribute to a great man and his vision of a better tomorrow.

But the development was a struggle from the beginning, and it took a toll on yet another member of the Disney family. Walt Disney World debuted to the public on October 1, 1971. Ten weeks later, Roy O. Disney died. Perhaps fittingly, the perception is that Disney's Parks and Resorts division suffered a void for 35 years after the death of the two titans who founded the company.

What's left in their absence is a strangely paradoxical tourist destination. Walt Disney World hosts the most popular theme park in the world, Magic Kingdom, as well as the sixth most popular, Epcot. The latter gate is the more notable in that it represents the best ideas that the Imagineers from Walt Disney's team could construct in the absence of their leader and founder. Its history reflects the difficulties that any company would face when the architect of a blueprint dies before construction can begin.

Magic Kingdom is an unquestioned masterpiece. Epcot, Hollywood Studios, Animal Kingdom, and the shopping district now known as Disney Springs fall into a different category. They are wonderful parks full of amazing, authentic themed lands. They are also all compromised, many would say damaged, products that failed to live up to their original promise. Whereas most of the headline-grabbing struggles at Disneyland came on opening day, the Walt Disney World complex has faced a constant spate of negative press due to its track record of overpromising but under-delivering.

I view these criticisms as too harsh. The blue sky phase of development is always the time when anything is possible. The constant constraints on resources always lead to lesser implementations of sweeping ideas. Even though Animal Kingdom and Hollywood Studios are undeniably half-parks that desperately needed the upgrades of Pandora: The World of Avatar, Toy Story Land, and Star Wars Land, they're still shocking feats of Imagineering.

Even Epcot, for all of its struggles in trying to live up to the dream of Walt Disney, is a triumph of innovation. Importantly, the park also offers the one thing that was most important to the company's founder. The World Showcase has succeeded as a daily World's Fair, just as Walt had desired in the wake of his company's triumph at the 1964 New York World's Fair. Sometimes, people forget that the most important aspect of a development is to get the big stuff right. That's exactly what Disney did with Walt's new world, the place that he didn't live long enough to see.

The following stories chronicle the many difficulties that The Walt Disney Company battled in building the most popular theme parks in the world. Each chapter demonstrates the difficult decisions that park planners faced in building thousands of acres of recreational activity. My hope is that you will read each one from the unbiased perspective of an Imagineer, weighing which path you would have taken if you'd faced the same decisions.

And remember: if constructing the Most Magical Place on Earth were easy, anyone could do it. In truth, only members of the Disney family and the loyal staff members that they trained were capable of accomplishing such a majestic feat.

CHAPTER ONE
Disney Landmarks

When you think about Walt Disney World, what are the first thoughts that enter your mind? Do you reminisce about walking down Main Street for the first time, seeing Cinderella Castle perfectly centered ahead of you? How about walking into Space Mountain, or the equally legendary Splash Mountain and Big Thunder Mountain? Do you remember that moment you were swept away by the majesty of Spaceship Earth at Epcot?

The scenery at Walt Disney World is an integral part of its appeal. The four parks claim a stunning number of architectural triumphs. Disney Imagineers often do the ultimate by building something from nothing. Many of their achievements are unforgettable, engineering marvels so impressive that college students research them for school papers. Have you ever wondered what was involved in getting them off the ground?

The short answer is millions of pounds of steel and other raw materials. The long answer involves a die-hard bunch of specially trained engineers with a can-do spirit and the boldness to succeed in ventures others wouldn't even have thought to try. That's the power of Imagineering. Of course, even Disney Imagineers face unpleasant realities such as budget constraints. That was particularly true during the early days of Walt Disney World, when building one attraction right regularly meant sacrificing another as the opportunity cost.

Countless landmarks at Walt Disney World are worthy of examination. Walt Disney's laser-focus on architectural design was something he passed down to all his disciples, allowing his legacy to last half a century beyond his death (and counting). Still, some of the achievements are greater than others. Let's take a closer look at the construction of several of the most awe-inspiring monuments at Walt Disney World, learning about all the struggles Disney employees faced in creating something from nothing.

Cinderella Castle

No other landmark in the world is as inexorably linked to its theme park as Cinderella Castle is to Magic Kingdom. To wit, the building is so beloved worldwide that Disney effectively duplicated it as the central point at Tokyo Disney Resort. Cinderella Castle is a 189-feet tall building including its watery moat. Its size is specific. Disney gained a lot of influence with the state of Florida during the building of the park, but they couldn't avoid FAA regulations involving structures in excess of 200 feet. Had Cinderella Castle stood that tall, government officials would have required Disney to add flashing warning lights for low-flying aircrafts. Imagineers weren't going to do that, of course.

While Disneyland fans in Anaheim maintain that Sleeping Beauty Castle is the original castle and therefore the greatest, Disney World park planners accounted for this thought process. During the crafting of the blueprints for Cinderella Castle, everyone involved understood the stakes. To branch out from Disney as a single theme park, they'd have to overcome the perception that the Florida park was derivative. The best way to achieve that goal was obvious. They'd have to make everything at Magic Kingdom bigger and better. When it came time to design the castle, they took that literally.

Cinderella Castle is more than 100 feet taller than Sleeping Beauty Castle. Thanks to the magic of forced perspective, it seems even bigger than that. People walking down Main Street may not even recognize that it's happening, but the castle will grow larger as they approach it. There's a trade-off, of course. The castle will also seem thinner, since that's how Imagineers intended the optical illusion of a regal skyscraper to work. Guests should always feel small in the presence of Cinderella Castle.

Bringing their planned landmark to fruition wasn't easy. If we exclude the infamous land purchases that Walt Disney managed to sneak under the noses of Florida real estate magnates, the company invested $4.7 million into the construction of Cinderella Castle, not including decorating the interior. You can look at that investment expense as a lot or a little. In 2015 dollars, that's $28.7 million. So, the castle cost less than Alex Rodriguez receives to play for the New York Yankees. The $4.7 million is also only a fraction of the overall cost to build Magic Kingdom. The total bill for the park was $400 million, which means Cinderella Castle is only one percent of the overall financial outlay. That's not too bad for a royal dwelling.

The primary concern with building Cinderella Castle was the time frame. Once Disney execs finished the blueprints, they were only 22 months away from opening the park. It would have been unforgivable for the central view of Main Street to remain under construction after the park had opened. The struggle everyone faced is that they broke ground on Walt Disney World in 1967. The new landmark was way behind some of the other projects.

Construction employees worked under a tight deadline. To their credit, they finished earlier than projected. Walt Disney World's Magic Kingdom opened its gates for the first time on October 1, 1971. Despite breaking ground for Cinderella Castle only 18 months prior, the project came together quickly. That's the benefit of great planning. It also helps that the backbone of Cinderella Castle is fabricated metal rather than brick, which is slower to set.

Six hundred tons of steel comprise the central spine of the castle. In order to secure it during hurricane season, there's also a reinforced concrete buttress that surrounds all the integral parts of the structure. Due to its cleverly engineered core, Cinderella Castle can withstand 110 mile-per-hour winds, and architects believe it could even go quite a bit beyond that if needed. Now that the architecture lecture is over, let's talk about the fun stuff. Imagineers modeled Cinderella Castle after several great palaces, fictional and real. Herb Ryman earned the right to construct the facility. He won rave reviews for his work in bringing Sleeping Beauty Castle into existence, so when Disney started to design their second theme park, he was the perfect choice.

Ryman chose some of the most beautiful buildings in the world as models. Those include Versailles and Fontainebleau in France as well as Moszna Castle in Poland. He also mimicked some of the ideas from famous châteaux such as Chambord, Chaumont, and Pierrefonds. The most lasting impression, however, came from the castle animated in Cinderella, as it should be.

Ryman requested ornamental accoutrements such as turrets and spires to emphasize the royal nature of the castle. To create the unusual look, he planned for Cinderella Castle to feature 29 towers. Two of those had to be scrapped because other plans involving the Fantasyland section of the park completely obstructed their view. For that reason, there are now 27 towers strangely numbered from 1-29. The two numbers for which there isn't a matching tower are 13 and 17.

Three of the most memorable towers are 10, 20, and 23. Tower 10 features the famous clock, 20 is the tallest of the group, and 23 is the unforgettable gold roof that you probably have appreciated in some of your favorite Walt Disney World pictures. Since the castle is supposed to be a place of tremendous romance, there's also a wishing well and bushes that Disney shaped specifically to emphasize that love is always in the air at Cinderella Castle. The romance is painted on the walls, too. The five murals in the castle archway reveal the love affair between Cinderella and Prince Charming.

There is one odd tidbit about Cinderella Castle. Disney emphasized authenticity from start to finish, even employing gold paint to enhance the regal perception of the structure. There was one aspect of Sleeping Beauty Castle that they couldn't translate to Walt Disney World, though. The drawbridge at Disneyland is functional. Disney has raised it exactly twice. The first time was for the park's grand opening, and the most recent instance occurred in 1983 when the company rededicated Fantasyland. By 1970, Imagineers understood that there was no point in designing a functional moat after Disney had opened its predecessor exactly once in 15 years. So, the drawbridge at Cinderella Castle is only for show. When the zombie apocalypse comes, you've been warned not to try to use it for safety.

Despite the storied history of Cinderella Castle, Disney isn't afraid to alter the look of its signature landmark. In 2015, the company updated the central plaza area with new gardens and pathways as well as a special viewing spot for Wishes. Since that show and Cinderella Castle are tethered together, Disney also started to update the castle. Construction workers began to build turrets. In July of 2015, two of these new structures finally came into view when cast members lowered the barriers hiding them. Two more debuted by the end of 2015. By expanding the space surrounding Cinderella Castle, guests received the benefit of more walking room. More important, the new touches accentuate the majesty of the castle that much more.

The Mountain(s)

In the years following the opening of Magic Kingdom, Disney faced tremendous economic constraints. For a few years, their architectural feats stood on hold. There was a reason for this status quo. While the new park proved wildly popular from the start, they still had to contend with the aforementioned $400 million cost to build it, the equivalent of almost $2.5 billion today. To say that they were cash-strapped undersells the problem.

Despite their financial limitations, The Walt Disney Company refused to let their founder's legacy flounder. They proceeded with several new projects over time. The catch was they faced opportunity cost decisions with each new construction. Big Thunder Mountain wound up the temporary loser in this fight.

Originally conceived in the early 1970s, Disney Imagineers intended the ride to operate as a companion to a linchpin attraction, which was already well into the planning stage. The Western River Expedition was to be a boating journey in Frontierland that would share the same general space as a train-style roller coaster. The company would craft a western-themed pavilion hosting the two new attractions. You've likely never heard of Western River Expedition since Disney never built it. In fact, that's where the entire conversation jumps the shark.

Imagineers projected Western River Expedition as the Pirates of the Caribbean of Walt Disney World. They saw no need to build a mirror copy of the signature Disneyland attraction in Florida, because that state was chock full of pirate history already. It'd be like building Hall of Presidents in Washington, D.C. They've lived it. Why do they need to catch the highlights again during a 10-minute boat ride?

Disney's logic was sound, but their branding was stronger than their logic. From opening day forward, Walt Disney World guests questioned when Magic Kingdom would add Pirates of the Caribbean. The only problem was that Disney hadn't planned for it. That means they also hadn't budgeted for it. In order for the company to give the people what they wanted/demanded, Imagineers would have to make sacrifices to build a ride they hadn't expected Floridians to anticipate.

The first cut was the deepest. Western River Expedition, the spiritual successor and Florida response to Pirates of the Caribbean, hit the chopping block. In order to lower expenses, Disney had to drop the pricey Adventureland expansion. Imagineers still liked the idea of a train functioning as a roller coaster, but again, there was a financing issue.

The company was no longer getting two attractions from the same financial outlay, and they'd experienced a sudden expense in building Pirates. Something had to give, and that victim was Big Thunder Mountain. The decision was only supposed to be temporary, though. Imagineers like Tony Baxter and Bill Watkins remained steadfast in their belief that this coaster would prove popular. Disney simply had to find the money to make it. In their estimation, that should have occurred as soon as Pirates of the Caribbean was ready. Except... Pirates of the Caribbean debuted to tremendous acclaim on December 15, 1973. With that attraction finished and drawing more customers into Walt Disney World, the company's financial problems eased a bit. Disney mastered the art of using increased ticket sales from one ride to pay for the next offering. Insiders presumed that Big Thunder Mountain would become that attraction. Only, it didn't.

Almost a decade earlier, vaunted Disney park constructor John Hench enjoyed a sudden burst of creative inspiration. He envisioned a space port that would become the hub of Tomorrowland. The central selling point of his new pavilion was an indoor roller coaster, the ultimate dark ride. Hench's ideas for something he called Space Voyage largely remained intact over the years as Disney planners tinkered with the mechanics of it.

In June of 1966, they changed the name of the ride to something more familiar, Space Mountain. Disney hired one of their favorite construction partners, Arrow Development, and they prepared to add it at Disneyland. Remember that Walt Disney World was still more than five years away when Space Mountain received its ultimate identity.

As was the case with so many attractions, Walt Disney's death fundamentally altered the history of Space Mountain. WED Enterprises wanted to carry on Walt's legacy, but their funds were limited. Investors were understandably skittish about the company after its founder had died. It's akin to Apple in the wake of the passing of Steve Jobs except that Disney wasn't the force of nature on Wall Street that Apple is today.

On some level, investors had confidence that Disney trained a worthy batch of successors, yet they didn't put their money where their mouths were when it came to loans. In the wake of Walt's death, the company found itself in a strange position. The stock increased steadily for years to the point that they become a member of the exclusive Nifty Fifth, indicating that they were one of the blue bloods in corporate America. Still, creditors worried that the debt they were accruing with their Florida purchases represented far too much unwelcome risk.

With a hard cap on their spending, Disney directed all their resources toward Walt Disney World. From their perspective, Disneyland was a dream that Walt lived to see come true. While they would always work to keep it fresh, a mission statement that still stands today, the better way to honor their leader would be to create a second theme park in Florida. They wanted to craft something just as impressive in the early 1970s as they had done in California in 1955. You see where I'm going with this.

Space Mountain, a Disneyland project, changed overnight into a Walt Disney World project. This happened in the wake of Pirates of the Caribbean, a Disneyland project, becoming a Walt Disney World attraction due to the demands of the Florida visitors. Big Thunder Mountain, a Walt Disney World project from the start, once again got pushed down further in the queue in order to make room for a California attraction nobody had originally planned to build in Florida.

From its original conception, Space Mountain was intended to be a modernized version of the Matterhorn Bobsleds. One of the reasons Disney hired Arrow in the first place was that they'd worked on the Bobsleds. The plans for Space Mountain were ambitious. It would contain four tracks that could depart the gate at the same time or at set intervals. Disney's financial issues as well as the space limitations expected for space port at Tomorrowland eliminated the multi-track option. For this reason, the premise of Space Mountain briefly fell out of favor with park planners.

Rather than create a new ride, the company realized they could save money by duplicating Matterhorn Bobsleds at Walt Disney World. Hey, it'd worked with Pirates of the Caribbean, and thrill rides were proving especially popular with the insatiable Florida crowd. Eventually, Disney execs recognized that they were in danger of Walt Disney World becoming known as Disneyland East rather than having its own distinct identity. They returned to the original plan, which was a new and improved coaster concept rather than a throwback to one from 1955.

In order to make Space Mountain stand out as something especially unforgettable, Imagineers appreciated that they'd have to do something different. That drive to build something novel led to the man-made structure so closely associated with Space Mountain today. Disney knew that the one advantage of building it at Walt Disney World was that they'd have a great deal more space. Tomorrowland was relatively barren in its early days. The problem remained budgetary in nature. One of Disney's top executives found a clever solution, enticing RCA to sponsor the ride in exchange for a $10 million investment. That was enough for Disney to feel comfortable greenlighting the $24 million project, out of which they would pay $14 million.

The exterior structure required a massive facility. Space Mountain would host the world's first indoor space-themed roller coaster. To get the details right, the interior roofing would project the Earth's place among the stars, thereby fostering the sensation of actual space travel. Gordon Cooper, one of the original seven NASA astronauts from Project Mercury, worked as a special advisor to the project to lend an air of authenticity. Award-winning author Ray Bradbury also participated, since science fiction is oftentimes more stimulating than science fact.

The two men collaborated with Disney illustrators to build a blueprint for Space Mountain the building. They designed the amazing 183-foot structure, which fell only six feet short of matching Cinderella Castle in height. Space Mountain features a circular design that was hotly contested for a time. Some Imagineers preferred a dome shape while others favored a cone. The latter option eventually won.

To bring the blueprints to life, Disney needed 4,000 giant pieces of steel plus 12,000 feet of electrical wire. The project required less than two years to complete once Disney finally broke ground, although the opening date of January 15, 1975, was over a decade after the initial conceptualization and pitch.

After the Walt Disney World version of Space Mountain debuted, Disney immediately focused on transferring the ride back to Disneyland. On May 27, 1977, the California version opened. Between the two attractions, the company invested more than a million man-hours on design and construction. The Disneyland version cost $4 million less at $20 million.

Perhaps because they were fresh out of excuses, Disney finally moved forward on the long-gestating train roller-coaster project once the Space Mountain build ended at Walt Disney World. Imagineers speculate that the delay turned out to be a net positive for the attraction. Computers evolved at an exponential rate during the 1970s. What was a nascent technology back when Tony Baxter invented the roller-coaster train concept had advanced dramatically over the course of the decade.

Of course, there was still a Charlie Brown trying to kick the football element to the construction. Amusingly, Disney execs once again flipped the placement of Big Thunder Mountain Railroad. After years of discussion about its inevitable location at Magic Kingdom, Disney *finally* broke ground...at Disneyland. You're likely noticing a pattern here.

The company played a shell game with all of their major projects during the 1970s. They followed a strict cost projection strategy based on where new rides would cost the least while bringing in the most new business. For whatever reason, Big Thunder Mountain Railroad at Walt Disney World lost this game so much that they would have picked #1 overall in the Phantom Disney Attractions Draft.

In September of 1979, the great train adventure finally debuted at Disneyland. Around the same time,

Disney determined that they should host the ride at both locations a la Pirates of the Caribbean and Space Mountain. So, nearly a decade after its initial proposal, the last remnant of Western River Expedition finally came to fruition at Walt Disney World.

While not the skyscraper height of some of the other attractions listed here, Big Thunder Mountain Railroad still stood 104 feet tall, more than Sleeping Beauty Castle at Disneyland. It again takes advantage of the ample Disney property in Florida, running over two-and-a-half acres of track, 25 percent more than its California counterpart. It includes a heavily sloped hillside that allows the coaster to turn at steep 270-degree angles while still delivering the experience of a runaway train during inclement weather.

Big Thunder Mountain's construction highlighted the conceit that a mid-19th century town had to be abandoned quickly. The end result left the mine trains somehow operating on their own. To reinforce the theme, Imagineers painted the mountain a reddish tinge to remind observers of the red rocks prevalent in the American southwest, one of the primary locations of gold rush settlers.

The next time you look at the mountain range, pay particular attention to the craftsmanship and attention to detail Disney demonstrated in crafting an artificial mountain so lifelike and vivid. Even as it stands next to another manmade façade, Splash Mountain, it still towers above that 87-foot tall structure. The twin towers have stood side by side since 1992, and it's difficult to separate the two when thinking of Frontierland. Had fate not been so capricious with Big Thunder Mountain Railroad, however, it could have been connected to a much different boat ride instead. I think we'd all agree that the situation worked out for the best.

Spaceship Earth

The Experimental Prototype Community of Tomorrow has yet to become a reality, and even if it does, it won't reside at the location known as Epcot. Still, Walt's vision proved integral to the construction of the park he imagined but didn't live to visit in person. And the signature site at the park is an 18-story tall geodesic dome that looks like a...well, it's...

In the early 1980s, giant steel balls were all the rage in society. Okay, there were at least two under construction. One was in Knoxville, Tennessee, where some of the finest engineering minds in the world readied an absurd looking hexagonal steel ball for the 1982 World's Fair. It was called the Sunsphere, and it's a punchline inasmuch as a lasting legacy to architecture. People quickly realized that using gold dust to layer glass panes would cause intense heating issues inside the building. That structure debuted in May of 1982 and was already a laughingstock by the end of the year.

A few months later, on October 1, 1982, EPCOT Center (as it was originally known) opened to the public. The jaw-dropping piece of architecture called Spaceship Earth stood mightily near the front gates, in sight of all the people ready to storm the park and enjoy its futuristic sciences and technologies. Suffice to say that this giant ball has stood the test of time a bit more easily than the earlier Sunsphere despite their similar themes. In other words, placing some of the finest engineering minds in the world against Disney's Imagineers proved to be a blowout win for Disney.

The key, as always, is that Disney employees anticipated concerns that people without experience in novelty engineering couldn't have expected. They debated many structural builds, each of which was capable of dazzling onlookers, before settling on one that wasn't just breathtaking to view, but was also feasible to construct and easy to utilize as a ride attraction. It's that kind of forward-thinking that explains why Spaceship Earth is as remarkable today as it was over 35 years ago when construction began.

While you'd need the world's largest driver to strike this particular golf ball, and only Godzilla would be large enough to swing it, there's no disputing the look of Spaceship Earth. It looks like a giant golf ball complete with dimples. Like so many triumphs of aesthetics, it looks completely different up close than it does from afar. That's because of the unique infrastructure.

The giant ball you know as Spaceship Earth has a geometric basis for its design, and one that sounds like a James Bond villain at that. It's called a pentakis dodecahedron, which means that it's got a series of 12-faced regions. A set of five triangular faces, each of them in a pyramid design, pairs with each dodecahedron.

Disney Imagineers modified the concept slightly to create the novel look you know today. Their giant ball includes 60 isosceles triangle faces split into 16 smaller triangles. Due to all the splits and cuts, we wind up with just under 1,000 flat panels (960 to be precise) totaling 11,520 triangles. All of them meet at 3,840 points. You can't see some of them due to side concerns for the building such as entry points. Then again, it's not like you were going to count all the way up to 11,000 anyway.

Don't worry. You're not going to be tested on any of this. It simply underscores how the complex underlying design of Spaceship Earth. Think of the entire structure as one giant SAT exam with a focus on your least favorite subject, geometry. If you find Hidden Mickeys searches obsessive, discovering all the triangles in the Spaceship Earth façade might drive you insane.

Now that you feel like a teacher's pet for knowing all the math of Spaceship Earth, let's focus on the

DISNEY LANDMARKS

construction. For all its complicated calculus, the actual building of the facility didn't take a lot of time. Disney completed the entire process in just over two years. The 26-month endeavor is a remarkable turnaround time for something that stands 182 feet tall and 165 feet wide. Note that Disney didn't feel the need to best its own records. They intentionally left Spaceship Earth seven feet shorter than Cinderella Castle.

What did Disney need in terms of raw materials to bring Spaceship Earth into reality? Well, it's more than you'll find at your local Home Depot. The theme of Epcot during its early days was "bringing the world together through technology," but raw materials must have been implied. Massive amounts of aluminum, plastic, and steel comprise the core. The design of the building is so strange that Disney had to build it in three parts.

The three sections were the sphere portion, the utility portion, and the ride itself. The ride and show portion had to fit inside the sphere part, which added complexity to an otherwise seamless process. For the ride, Disney built a floor unit with concrete slabs supported by metal decking. The utility portion does most of the heavy lifting in a literal sense. It handles the load transfers for both the sphere and the ride, thus the name utility.

Due to the novel design of a geodesic sphere, Disney had to research its feasibility. They were especially worried about the weather conditions in Orlando. Disney asked MIT to perform wind test studies on a smaller scale version of Spaceship Earth specifically built for testing purposes. Disney employees needed to make sure that guests walking around the base of the structure wouldn't have anything ridiculous happen. There were early concerns that wind might knock guests off their feet. Thankfully, those worries proved unfounded.

Disney still wasn't done worrying about the weather, though. Imagineers calculated the potential flooding issues, particularly during hurricane season. It's true that if you stand directly under Spaceship Earth, you won't get wet. This is a feature, not a bug. One of the hidden purposes of the many, many triangles is that they drain water in a way that shields guests from moisture. Spaceship Earth is more than an architectural triumph. It's also a giant umbrella.

Finally, in order to create the floating effect of the ball, Disney put it on stilts. Yes, I'm serious. The actual structure stands 18 feet above the ground. That's because a series of six steel legs provide its balance. They are built 160 feet into the ground to provide maximum support for a facility that weighs an almost incomprehensible 7,760 tons. That's the equivalent of 130 elephants. Eat your heart out, Animal Kingdom! In the era of multi-billion dollar purchases such as Pixar and Star Wars, we sometimes forget that Disney once faced significant budgetary constraints. Even after they took on $400 million in debt, they continued to find ways to increase the value of their theme parks for visiting guests. Many of the architectural triumphs we take for granted today wouldn't have been possible for even the best designers of that era, even if they'd had limitless budgets. Disney's Imagineers continually turned seeming disadvantages such as limited financing, governmental regulations, and lack of raw materials into long-term advantages. Rather than be dissuaded from their ultimate goal of building these majestic attractions, they simply prioritized the ones that were more feasible in the short term. Then, once they had more money and benefited from technological advances, they crafted better rides in the same locations. When you visit Walt Disney World, every architectural accomplishment is also a monument to the ingenuity and perseverance of Disney Imagineers.

CHAPTER TWO
World Showcase

You know the story by now. Walt Disney's vision for his Experimental Prototype City of Tomorrow included a constant World's Fair site. The founder of what we now know as The Walt Disney Company enjoyed his finest moment at the 1964 New York World's Fair. It was during this two-year global event that the Imagineers of WED Enterprises cemented their reputation as some of the finest engineers in the world.

Disney built four attractions for the World's Fair, and then they persuaded other companies to pay for their transportation to new permanent residences at Disneyland. One of them, Carousel of Progress, still exists (mostly) in its original form today at Walt Disney World. Another, It's a Small World, has become one of the most famous theme park attractions on the planet.

As much as Disney as a company influenced the World's Fair, however, the event itself similarly left a lingering imprint on Walt. He felt a profound connection to the world travelers who all came together in New York City, and he was touched by how much they loved his new creations. From that moment forward, the head of the Imagineers aspired to create a permanent host site for the people of the not-so-small world.

The method for bringing this fluctuating but constant World's Fair into existence eventually became the World Showcase at EPCOT Center. By the time it arrived, however, the company, and the entire world, for that matter, had lost the visionary who'd conceptualized its existence.

In their founder's absence, Imagineers had to soldier on, doing the best job that they could of turning his wishes into reality. The problem they faced was that Walt Disney as a businessman was every bit as gifted as Walt Disney as an inventor or storyteller. Without his power and influence, many sacrifices were necessary to finish EPCOT Center.

Perhaps the loss of the man who "invented" Walt Disney World explains why EPCOT Center took so long to construct and included so many attractions that aged poorly. Whatever the explanation, park planners did cross off one key component from Walt's checklist. Half of the new park functioned as an ever-present World's Fair, a host site for pavilions from many different countries.

In the early days, the World Showcase portion of EPCOT Center had grand ambitions as an often-expanding cultural melting point with a finger on the pulse of global interaction. Over time, those dreams became convoluted by a messy combination of politics and money. While nobody realized it at the time, the international representatives of Norway would introduce the final pavilion in 1988. Meanwhile, promised appearances from Africa, Israel, Poland, Russia, Spain, Switzerland, and Venezuela never materialized.

What's the explanation for the awkward, unchanging composition of the World Showcase? How did Disney officials choose the nine original countries and two later additions? What did they view as the primary appeal of each one? And what went so wrong that Disney stopped building pavilions in spite of promises and press releases stating otherwise? Everyone has a strong opinion about the relevance of Future World in the Age of Technology, but don't many of those same criticisms apply to World Showcase? Let's find out...

"This concept here will have to be something that is unique, so there is a distinction between Disneyland in California and whatever Disney does in Florida."

"And EPCOT will always be a showcase to the world for the ingenuity and imagination of American free enterprise."

These quotes are attributable to Walt Disney. He recited each one during the press phase of the EPCOT announcement, and each reflects the excitement and optimism he felt about the new endeavor. Sadly, Walt passed away in 1966, only two months after the company unveiled the Florida Project film. That left his amazing staff of Imagineers with an impossible situation. They had to anticipate what the company owner had intended for the 27,000 acres of land Disney had stealthily acquired from unassuming Floridians. Their short-term financial loss was beneficial to the future of central Florida as a whole. Walt Disney World eventually became the most popular set of theme parks in the world, but their early struggles are well chronicled by now.

In order to finance the other parts of Walt Disney's plan, his brother Roy O. Disney had to focus on monetization early on. Purchasing the land had left the company strapped for cash, and they'd already adapted their sweeping goals for the utopian city even before Walt's death. Both men understood that theme park ticket sales would pay for the other aspects of the City of Tomorrow. Still, no one expected an 11-year gap between the debut of Orlando's Magic Kingdom and EPCOT Center.

The explanation for this delay is that Walt's death left his successors without a guided path forward. They knew of his vague plans for a permanent World's Fair site. Many of them had witnessed the 1964 New York exhibition firsthand, and they recounted stories of their employer's joy over the reception of the four Disney pavilions. They understood that Walt Disney World would include something along those lines, but it's akin to your boss telling you that you need to complete that project you discussed last year. You lack the specifics to feel comfortable about the project's direction without additional input. With Walt Disney gone, nobody would offer that counsel. Imagineers had to make it up as they went along.

Somehow, all parties worked together well enough to launch a version of EPCOT Center worthy of Walt Disney's legacy. A great deal of credit should go to the employees involved as well as the man who trained them as Imagineers. They embraced the challenge and unified under a singular purpose: honoring their leader's vision of a better world. A key component of that new and improved tomorrow was an international hub. The politics of it were a nightmare, though.

Unbridled enthusiasm embodied the earliest days of planning EPCOT Center. The launch of Magic Kingdom left Imagineers and park planners in a celebratory mood. The park had instantly become one of the top tourist destinations in America. While the company still suffered financial challenges that forced them to choose selectively among options for new attractions, everyone involved felt that the first Walt Disney World gate had succeeded beyond all reasonable expectations. Once the corporate purse strings loosened, they had complete confidence in phase two, the place we now know as Epcot.

The planning phase of the second Florida gate came from strategists brimming with confidence. They saw no reason why countries would choose not to join Disney in their unprecedented endeavor. In their view, the new locale would offer its participants the opportunity to grow the brand, so to speak. For a country, that meant indoctrinating tourists to a new culture brimming with authentic foods, merchandise, and artisans. The first members of the World Showcase would effectively gain a new point of ingress to American dollars.

And here's where the story takes a turn. To bring the truest version of Walt Disney's dream to life, park planners reached out to 31 countries. Their sales pitch was simple. Disney would build two parks in one space, the equivalent of a constant World's Fair site. The front half would highlight emerging technologies while prophesying future feats of science and engineering. It would mimic Disney's 1964 pavilions such as Ford's Magic Skyway. This portion would attract adults seeking something a bit more mature than the Mickey Mouse-based fare available at Magic Kingdom.

The back half of EPCOT Center, the World Showcase, would cater to the same guests in a different way. It would feature permanent shops and restaurants inside breathtaking facilities representative of the presenting countries. Each participant would enjoy input into the design of their pavilion, a first for Disney as a company. They contracted some engineering work to businesses such as Arrow Dynamics, but Disney had always maintained control of the creative process. With the World Showcase, they were willing to cede control of some aspects to the governments who knew their heritage and aesthetics better. It was a thoughtful gesture on Disney's part.

How many of the 31 countries with Disney offers would you guess signed up?

Unless you guessed zero, you're wrong.

Amazing as this proposition is to consider, no country felt the onus to join Disney's new enterprise. Their governments offered a multitude of reasons to explain the universal rejections. Some parties worried that by joining an unsanctioned institution masquerading as a de facto World's Fair, they might incur the wrath of the International Exhibitions Bureau (BIE), the governing body of official World's Fairs.

While the idea seems quaint today, losing World's Fair privileges caused a great deal of consternation during the second half of the 20th century. Amusingly, the 1964 New York World's Fair faced the same issues since it too was unsanctioned. The bylaws of the BIE stated that such events could only last for a year. New York's event actually ran for parts of 1964 and 1965. Meanwhile, the World Showcase at EPCOT Center would remain open constantly. That was one of the core concepts of the facility. The BIE rattled its saber about severe punishment for countries who stepped out of line. At the time, their threats had some teeth.

Bureaucracy of that sort was only a small part of the problem, though. Actual political bureaucracy was a larger issue. Every potential country needed to garner the favor of a majority of their elected officials or in some cases that of royalty or unelected rulers. That task is never easy in any day or age.

Even when governing bodies agreed, they also needed to form alliances with popular businesses from their regions. Disney emphasized commercialization prior to the launch of the World Showcase. Without major corporations on board, the stores wouldn't seem aptly reflective of the countries they represented. While commerce was king, the negotiations also required a political stamp of approval. Imagine trying to get the American government to agree to an endeavor but also needing the approval of major corporations like Google, Facebook, and Apple. That's far too many conflicting interests to bring together at once.

Finally, Disney just couldn't win the public relations battle. While we all know and appreciate the legacy and import of Epcot today, foreign governments of the mid-1970s didn't understand the sales pitch well enough to waste political capital supporting it. Disney as a brand was already unimaginably popular worldwide, but EPCOT Center wasn't selling Mickey Mouse at all. To the contrary, it was stubbornly standing apart from that image, instead highlighting international cultures. The potential countries didn't understand why they needed that type of PR in what was up until very recently Florida swampland.

In 1975, Disney delivered a message to shareholders that brimmed with optimism. It highlighted all the eventual wonders of the World Showcase at EPCOT Center. Park planners believed that "the World Showcase will communicate the culture, heritage, history, technology, trade, tourism and future goals of the participating nations." More than 40 years later, they're largely right about what transpired. It was the getting there that caused the problems.

The shareholder message indicated a 1979 debut for the World Showcase. As you know, that was three years premature. And their original goals fell by the wayside. Disney announced:

> These dynamic structures will face each other across a Courtyard of Nations, where there will be an outdoor theater for performances by international celebrities and entertainment groups, and where parades, pageants and special events will be staged by entertainers from the participating nations.
>
> Although these national pavilions may vary in size, each will enjoy equal façade exposure to the guest. The entire complex will be tied together by a Disney people-moving system that will offer visitors a preview look into each attraction.
>
> Unlike a world's fair, it will offer participating countries a permanent installation for such

features as themed restaurants and shops, product exhibits, industrial displays, cultural presentations, a trade center, and even special facilities for business meetings. A major part of each pavilion will be a Disney-designed ride or attraction which will give guests a foretaste of an actual visit to the country. National musical groups or other performing artists could present special entertainment on a continuing basis.

As you flip through your mental rolodex about the validity of these claims, consider that Epcot does feature international celebrities and entertainment groups on a regular basis. They also have outdoor theater performances thanks to talented cast members. The Courtyard of Nations with "equal façade exposure" is also largely accurate. Many of the merchandising and cultural claims are also valid.

Where Disney loses points for prophecy is with the Disney-designed attractions. You can count on one hand the number of rides offered, and the situation only looks marginally better if you generously describe the outdated country-based films at some pavilions as attractions, too. And while that might sound nitpicky, it underlines where Disney miscalculated the interest from foreign parties.

A different part of Disney's anticipated sales pitch is where they ceded any chance of enticing countries and international businesses to participate. From Disney's perspective, they were building a permanent World's Fair. One of the presumed aspects of the pavilions at such exhibits was that the participants all paid their own way.

In Disney's estimation:

Each participating nation will be asked to provide the capital to cover the cost of designing, developing and constructing its attraction and/or ride and all exhibits, as well as the Pavilion itself. It will also have the responsibility for funding the housing for its employees in the International Village. Its land lease will cover the cost of maintaining the attraction for a minimum of ten years.

They felt this request was perfectly reasonable, as it was in line with standard World's Fair practices. The potential countries involved already didn't like the idea of upsetting the BIE. Paying to have a tentpole pavilion in central Florida made no sense to anyone. The Orlando of the mid-1970s wasn't what it is today, the unofficial theme park tourist capital of the planet. Countries were already shaky on the premise. The fact that it came with a rather significant price tag proved to be the deal breaker.

Disney could have persuaded some countries right from the start if they hadn't tried to stick these governments with the bill. That's the explanation for why none of the 31 countries said yes. Disney had to use different methods to persuade them, and even those weren't certain to work. As late as 1979, Disney warned shareholders that the World Showcase could open with only five nations as participants. That's not much of a World Showcase. In fact, it's barely half of the number of countries in Western Europe.

After a firm rebuke from 31 different nations, Disney park planners faced a seemingly impossible challenge. They had to bring the vision of their dead founder to life in a way that would honor his memory. And they had to do it without the passionate support of any single nation. Corporate execs felt confident that if they could entice a couple of major countries to join the World Showcase, others would feel more pressure to join.

Their logic actually proved sound by the end of the 1970s, although they'd have nothing to show from it in the most famous example I'll discuss momentarily. During the most troubling days of the World Showcase,

Disney couldn't even convince America's best friend, England, to commit. At this point, everyone involved accepted that they'd have to change their sales pitch.

A new outline for the World Showcase reflected the company's desperation. While noting the lack of participants at the start, Disney explained to its shareholders that companies simply weren't willing to spend money on a questionable tourist venture in a foreign country. For all they knew, it was the Nigerian Prince scam of the 1970s, respect for the Disney brand notwithstanding.

The World Showcase would still finance itself, though. Disney would target international businesses rather than governments. Park planners accepted that they'd aimed too high with their original outreach program. A handful of countries acknowledged that they were willing to participate in the World Showcase, even at the risk of incurring the wrath of the BIE. They would only do it for free, though.

The World Showcase needed a different revenue stream to survive. What park financiers started to do was reach out to potential countries. This new diplomacy mission simply required a head count of who would provide a definitive yes if Disney paid for everything.

Once they had those countries onboard, Disney reached out to some of the most prosperous corporations from said countries. They brokered deals to split revenue with any businesses willing to pay for some of the initial costs. As I've mentioned before, one of the recurring issues when Disney opens new theme parks is that they run into cash-flow issues.

The best way to solve such financial woes is by finding interested parties to agree to joint revenue plans. It's how Disney financed several of the 1964 New York World's Fair programs before persuading the corporations to pay for the transportation of those attractions to Disneyland. Walt would have loved the ensuing World Showcase negotiations since they aptly represented his business skills.

When EPCOT Center debuted in 1982, the World Showcase included nine pavilions, eight of which represented foreign countries. No international government paid as much as a dime for these pavilions. Disney still managed to offset part of the $800 million cost of EPCOT Center through partnership agreements with corporate entities. Here's the list:

- The United States Pavilion aka the American Adventure had sponsorship deals with American Express and Coca Cola. In lieu of mom and apple pie, those are the most obvious brands available. Disney has negotiated long-term deals with Coca Cola *and* Pepsi Cola over the years, too.

- The United Kingdom Pavilion included three licensing and sponsorship agreements with Bass Export Ltd. (now known as Bass Brewery), Pringle of Scotland (a knitwear retailer), and Royal Doulton (glassware and collectibles).

- The France Pavilion enjoyed the second largest number of sponsorships, one of which continues to pay dividends for foodies today. Barton & Guestier (winemakers), Guy Laroche (perfume and clothing manufacturers), Lanson Champagne, and the France Chefs all licensed deals with Disney. The France Chefs included Paul Bocuse, founder of the legendary cooking competition, the Bocuse d'Or. He and his family have also hosted one of Epcot's finest restaurants, Les Chefs de France, which his son Jérôme continues to operate today.

- The Japan Pavilion only managed one sponsor, but that vendor has proven incredibly lucrative over the years as well. Mitsukoshi, Inc. has run Mitsukoshi Department Store since the park's inception, and the retail outlet maintains its status as arguably the most popular merchandising location at Epcot to this day. Fans of Kawaii get stars in their eyes when they think of its merchandise.

- The Italy Pavilion had two sponsors, and they were of the food and wine variety, a tactic that later became Epcot's bread and butter. Alfredo, the Original of Rome aka L' Originale Alfredo di Roma Ristorante traced its lineage back to the old country. It was there that Alfredo Di Lelio founded a restaurant bearing his name that has become an Italian staple over the past century. Epcot's securing a central Floridian franchise was something of a coup, and they secured a 25-year contract. The restaurant remained open for exactly 25 years, as Alfredo refused to extend the deal. The other sponsor was Brolio/Ricasoli & Bersano Wines of Italy, the second oldest winery in the world.

- The Germany Pavilion claimed the most sponsors. Five of them signed up. Bahlsen is a 19th century food producer whose products are still available at Epcot today. Beck's Brewery, also known as Brauerei Beck and Co., provided the adult suds that made every day at the park feel like Oktoberfest. Wine also had representation thanks to Schmitt Söhne. The other authentic cultural tie-in disguised as a sponsorship was Goebel. Their store would sell the finest products, and artisans would occasionally personalize them for Disney tourists. While their products are still available, the store is now a Werther's. Hutschenreuther joined Goebel in

- selling products befitting German heritage with their line of Bavarian porcelain.
- The Mexico Pavilion also followed the food and wine path. Moctezuma Brewery was one of the most popular breweries in the world, eventually winding up as a purchase of Heineken International during the 21th century. The other sponsor was unique. It was the entire San Angel Inn construction, including the building as well as the stores and restaurants inside. It's effectively a theme park mirror of the very real property in Mexico City.

Neither Canada nor China had sponsors for their pavilions in 1982, showing that Disney wasn't always able to broker a deal. The fact that several of these sponsorships lasted for decades wasn't accidental, and it directly leads to Disney's expansion problems that followed. Let's discuss them now.

As early as 1979, Disney brokered deals with countries that still don't have a presence at the World Showcase. Oddly, some of those negotiations completed successfully, only to fall apart later. Sometimes, the blame was Disney's. On other occasions, the governments of the potentially joining countries changed or the corporations financially backing them backed out of a completed agreement.

In one specific instance, nobody could have foreseen the collapse of the deal. The implosion of the deal was... whose name would go first in a multi-country pavilion agreement. That's a real thing that happened and the perfect example of how much turmoil EPCOT Center park planners faced in putting together the World Showcase as we know it.

Perhaps we can learn the most about why certain pavilions failed from the ones that arrived after EPCOT

Center's debut. Once the World Showcase opened on October 1, 1982, Disney had proof of concept for the next wave of countries desiring their own pavilions.

Park planners could show the revenue stream available to corporations willing to sign sponsorship agreements. They could also use the current nine pavilions as proof that EPCOT Center would boost tourism for the participating countries. This thought process is the reason why Disney accepted that they didn't need financial backing from the original nine. They felt they could give away the first nine in exchange for paying members later. Once non-participating countries saw the benefits, Disney felt confident that these governments would more readily open their wallets.

Was Disney right?

I'm going to blur the timeline a bit here to discuss the first two additions to the World Showcase since they're instrumental to understanding what Disney sought in their negotiations. As you may know, Morocco became the tenth participant in 1984, less than two years after EPCOT Center opened. It was the first pavilion to accept the business arrangement that Disney execs desired.

King Hassan II of Morocco ruled his country from 1961 until his death in 1999. The son of the legendary Sultan Mohammed V, he and his father led Morocco for 90 years, a staggering statistic. Despite the longevity of their rule, King Hassan II was only a decade removed from a coup attempt by his minister of defense when EPCOT Center opened. As a way to secure his perception in western civilization, the Moroccan ruler decided to construct a pavilion at Walt Disney World.

From Disney's perspective, the human rights record of King Hassan II was problematic, but a Morocco Pavilion embodied Walt Disney's vision for a permanent World's Fair. Their culture and heritage is catnip to North Americans, and so they overlooked the potential political ramifications. It helped that the country's ruler had sided with the United States, negating the potential public outcry at the time.

Best of all, King Hassan II was a billionaire. People suspected this during the early 1980s, but it's since been confirmed that he was one of the richest men who ever lived. The Moroccan ruler willingly paid for the entirety of the Morocco Pavilion. He cared about the project in a way that Disney had always hoped every country would.

Hassan II sent dozens of his finest artisans to Orlando. Their sole purpose was to create a pavilion befitting the people of Morocco. The building itself features a minaret that would seem perfectly at home in Marrakesh. That's because they used a real one as the inspiration. The famous commerce of Morocco also received special attention at the pavilion. The thematic bazaar is the merchandising capital of the facility.

The current version of the Morocco Pavilion, still at least somewhat overseen by the Moroccan government, varies little from its original setup. That's a tribute to the reverence the Moroccan people displayed in providing westerners with an authentic replication of their culture. More than any other pavilion at Epcot, this one exemplifies what Disney wanted and expected from the World Showcase. Each pavilion should accurately display a country's heritage, and said country should pay for everything.

The Norway Pavilion has certainly received the most attention of any World Showcase participant in recent years. That's due to a little film called *Frozen*. The ramifications of the most popular animated movie of all-time still ripple through Epcot today, but the early days of this project went exactly as Disney had hoped for all participants.

The government of Norway felt that their nation suffered from a tourism problem, particularly with

North Americans. They carefully monitored the World Showcase during its early days and watched with more than curiosity when Morocco joined the showcase in 1984. At this point, Norway's deciders chose to jump in with both feet. History has justified their decision a thousand fold due to the *Frozen* factor, but it was already a great idea during the 1980s.

Due to the Norwegian government's desire to add their presence to the World Showcase, Disney brokered another strong deal. The people of Norway invested at least $33 million in an endeavor that cost…$33 million. The Mouse House once again added new selling points to their park without spending much money. No less than the crown prince of Norway backed the project. The unmistakable conclusion is that Disney eventually realized that royal governments were less mercurial than ones with constantly rotating rulers.

The Norway Pavilion was also close to what Disney park planners intended with the World Showcase. Since the Norwegian government couldn't fund the entire project, they sought donations from corporations within the country to secure the requisite funding. Disney wound up with an entire government and a consortium of corporations tethered to the pavilion. It was again the most beneficial possible arrangement for them.

By telling this story out of order, I've shown you exactly how the process looks from Disney's side. With the first 31 requests, they faced sound rejection. Every possible country swiped left. Then, when Disney offered to pay, nine countries expressed the marginal interest needed to proceed. After those pavilions all worked well, EPCOT Center officials enticed two other countries not just to participate but also to pay for everything.

Your confusion at this moment is understandable. If Disney could broker deals like this, why wouldn't they continue? No downside exists to a deal where one party boosts the appeal of their park while the other pays for everything. It was never that easy.

What I've detailed are the "success" stories, some of which required a great deal of Disney coercion and hefty financial outlays from the corporation. The other side of the equation involves the failed negotiations. A shocking number of these exist. Let's start with Israel, a country who actually signed a contract to join Epcot. Does that surprise you? Would you be shocked to learn that Disney sent out a press release trumpeting this agreement? It's absolutely true.

Here's the pertinent text from the 1981 statement:

> The Venezuelan pavilion joins a number of countries and areas of the world scheduled for addition to the eight inaugural World Showcase nations. They include: Mexico, China, Germany, Italy, Japan, France, the United Kingdom and Canada. Venezuela, the State of Israel and Equatorial Africa are among the Phase II pavilions planned.

That last sentence likely confused you so much that you had to read it again. It's not your imagination. All of those pavilions were close enough to construction that Disney felt ready to announce them to the press. None of them wound up breaking ground. What went wrong? A little bit of everything is the depressing answer.

Nervous members of Disney's board expressed concerns about Israel's presence. EPCOT Center was supposed to grow the Walt Disney World brand by encouraging even more tourists to head to central Florida. The presence of an Israel Pavilion sadly caused worry about terrorism. Even 35 years ago, park planners fretted over security concerns involving EPCOT Center, as did federal investigators.

While the rich history of the Israeli people seemed like a wonderful expansion of the World Showcase, park officials believed that the best-case scenario for the Israel Pavilion would involve frequent protests. Disney never officially updated the public on this venture, but their fears were later validated, at least somewhat, in 1999. The Arab League called for a boycott of Epcot due to its 18-month Millennium Village addition, which included a temporary Israel Pavilion.

As for Venezuela, they joined Spain on the list of countries with governments who longed to add a presence at EPCOT Center. Both countries paid the $500,000 required to build blueprints for a pavilion, a Disney requirement for potential Phase II participants. But their governments changed hands or lost influence and financial flexibility around this timeframe, causing current or new regimes to back away.

The heartbreak here is that the blueprints for these pavilions are available online. If you decide to look at them, you'll note that Venezuela would have featured an aerial tram to take guests through the rainforest. Spain's blueprints were even more ambitious. They included two attractions, which is double what any current pavilion possesses. Those aren't even the most tantalizing plans, though.

When Norway struggled to come up with the required $30 million, other countries from the region pounced on the idea of a pavilion representing several countries. Once Norway's corporations found the money, Switzerland's government plotted their own pavilion. They didn't want to cede tourism to Norway. One of the most exciting plans ever for the World Showcase almost became a reality. The Switzerland Pavilion would have included a Walt Disney World version of the Matterhorn.

Perhaps the strangest collapse of a deal is the one I referenced earlier. Disney announced an Equatorial Africa Pavilion in 1980. The Ivory Coast, Kenya, and Senegal all agreed to join forces for a joint pavilion.

Each one signed paperwork, and the economics of the agreement weren't an issue. The reason this pavilion doesn't exist today is infuriating.

The countries squabbled over whose name should go first on the various listings. All three of them would enjoy equal representation and a windfall of tourism from the agreement. Jingoism prevented that from happening. Unreasonable national pride led to all parties walking away, leaving Disney execs feeling bitter about the experience.

In fact, let's examine the situation from Disney's perspective. It'll enlighten you on why the World Showcase has remained static for the body of three decades. After originally going 0-for-31 in persuading countries to join, Disney finally got nine to play along…as long as Disney paid.

After these nations agreed, others changed their stance, at least slightly. Over the course of a decade, Disney entered negotiations with Costa Rica, Denmark, Iran, Israel, Ivory Coast, Kenya, Russia, Senegal, Spain, Switzerland, the United Arab Emirates, and Venezuela. Eleven countries currently have a presence at the World Showcase. That's *twelve* others Disney desperately worked to add to the park, failing in each instance. Out of those missing dozen, only Israel isn't there due to Disney's choice.

Why are there no new World Showcase pavilions as of 2017? Getting eleven was such a nightmare that it's completely understandable why Disney eventually just gave up.

CHAPTER THREE
Disney Spies

Communists are trying to invade every facet of American society! Even the Most Magical Place on Earth is vulnerable to commie infiltration. Won't somebody please think of the children?

Sure, the above sounds like the less-than-subtle message from a post-World War II propaganda film. Almost unbelievably, this line of thinking has a basis in fact. Years prior to the opening of EPCOT Center in Orlando, Florida, several different American intelligence gathering organizations carefully monitored the development of the second gate at Walt Disney World.

Before scoffing in disgust, you should think about the situation from a historical perspective. In the late 1970s, the Second Red Scare was roughly two decades old, which may seem like a lot until you consider it's basically the same timeframe from the opening of Disney's Animal Kingdom to today. To anybody over the age of 40, the massive media coverage of Animal Kingdom's debut doesn't seem like forever ago, simply an earlier point in life.

The same was true in the 1970s of the Second Red Scare as well as a Cold War that continued until Charlie Wilson came up with a plan. This explains why the continued fear of communism, socialism, and all things Russian remained a pervasive part of American culture well after World War II ended.

Until the Berlin Wall fell in 1989, all good-hearted citizens cast a wary eye toward Eastern Europe. Some of the most ardent supporters of the American way of life found a way to turn their patriotism into a professional calling. These well-intended folks joined secretive institutions such as the Federal Bureau of Investigation and the Central Intelligence Agency. Their primary goal was to protect the American way of life, one they believed that communists threatened.

What follows is the shocking story of how the theme park we now know as Epcot once stood as a potential cultural turning point for the United States. It was here that people feared our country was most vulnerable. This article will describe how and why Epcot reached this point as well as how many of the fears proved to be founded in truth. We'll evaluate the no longer redacted files that reveal how much espionage transpired just outside the Magic Kingdom as the opening of Epcot approached. And we'll also discuss how much the politics of Walt Disney and the company that continued after his death played into the chaos of the situation.

Walt Disney was more than just a proud American citizen. He was a war hero of a sort. No, he didn't fight in either World War I or World War II. It wasn't for lack of trying, though. He quit high school at the age of 16 specifically with the intent of becoming an Allied soldier. Since he was only 16, he wasn't allowed to join.

Eventually, Disney and a friend participated in the Red Cross, which greatly expanded its global presence during World War I. Due to the tremendous training involved, Disney wouldn't even serve as a Red Cross ambulance driver until the war had technically ended. What he learned during this tumultuous period was that the world was a more beautiful and challenging place than his homes in Kansas City and Chicago. After getting a taste of the world, Disney wanted to make his mark.

We all know that he did so in the field of animation. His animation abilities, his entrepreneurial skills, and his natural storytelling gifts helped him become successful quickly. Around his 30th birthday, Walt Disney was already responsible for the creation of the most iconic cartoon character in the world, Mickey Mouse.

By the end of the 1930s, people across North America revered Walt Disney. His seminal animated film, *Snow White and the Seven Dwarfs*, became so popular that Disney received an honorary Academy Award for its creation. When World War II unfolded, Walt was already in his 40s, too old to serve in a military role, but he wasn't about to sit on the sidelines while his country fought in such a significant worldwide conflict.

War propaganda films became his stock and trade as younger Americans fought against the Axis Powers overseas. The sheer volume of these films was partially a tribute to the marketing power of Walt Disney. His war movies underscored the importance of able men doing their part to stop Hitler.

Starting in 1942, only months after the bombing of Pearl Harbor, Disney's filmmaking team crafted edutainment ads for the United States Navy. Later, the Army and the Air Force asked for similar propaganda pieces. As the country experienced a greater need for materials and tax revenue, Disney added the Department of Agriculture and the United States Treasury to his growing list of war clients. In a fascinating trivia note, one of his contacts for the government was a colonel named Frank Capra. You know him better as the director of *It's a Wonderful Life*.

Disney's contributions to the war effort cannot be overstated. His crew shot an estimated 400,000 feet of footage. That's the equivalent of three full days of movies enticing Americans to do their part to help their fighting soldiers abroad. His credibility in all phases of society influenced people in ways that politicians and statesmen could not.

For example, one of Disney's Treasury Department films explained the consequences of many Americans cheating on their income taxes. During a time of war, the government needed as much money as possible. Disney petitioned viewers to pay their fair share to fill the economic gap. A Gallup survey performed in the wake of the war film indicated that roughly 30 percent of viewers of the propaganda piece either paid taxes for the first time or paid them earlier than normal in order to boost the United States' war chest.

Certain aspects of Disney's life were controversial at the time and even more so now. He was a diehard Republican and frequent fundraiser for conservative issues. Most of his political beliefs involved the need to protect America from the influx of socialism. After World War II ended, a perception existed that the United States and Russia were headed to war at some point. The only question was when.

From a historical perspective, all the countries Germany had conquered were susceptible to new influences. Men like Disney worried that Russia would steer these nations toward communism. The moment Hitler's reign ended and Emperor Hirohito surrendered on behalf of Japan, people on both sides of the Atlantic quickly soured on the alliance that saved the planet during World War II. Disney in particular remained wary of what would become the Soviet Union for the rest of his life.

One of the most unfortunate legacies of Walt Disney was his support of McCarthyism during the 1950s. He testified before the House Un-American Activities Committee. The gist of his commentary is that Disney's anti-socialist reputation was so established that other people within the industry orchestrated strikes and smear campaigns against Walt Disney Studios. This tactic only steeled Disney's resolve. Here's the testimony that clearly defines why he feared the growing socialist influence in Hollywood:

> Well, they distorted everything, they lied; there was no way you could ever counteract anything that they did; they formed picket lines in front of the theaters, and, well, they called my plant a sweat-shop, and that is not true, and anybody in Hollywood would prove it otherwise. They claimed things that were not true at all and there was no way you could fight back. It was not a labor problem at all because—I mean, I have never had labor trouble, and I think that would be backed up by anybody in Hollywood.

Walt's reputation took a hit across the globe, especially in regions where socialism enjoyed more popularity. He was the millionaire businessman taking advantage of the sweat of the working class, the proletariat so important to the underlying theory of Marxism. To communists across the planet, Disney was an extremely public enemy. That intense dislike was mutual, a rarity for a man otherwise chronicled for his kindness.

This is where the situation grows trickier, more nuanced. The company founder was unquestionably a political partisan. An argument could be made that this philosophy has either evolved or skipped a generation. Did you know that one of Disney's grand-nieces, Abigail Disney, is an accredited filmmaker in her own right? But her political beliefs skew heavily to the left left.

Abigail Disney reflects something fascinating about the politics of Disney. They're difficult to pin down accurately. On social issues, the company is above reproach. Their leadership in inclusive hiring practices is the gold standard in the corporate realm. They are thought leaders who back up their theories in practice, hiring people from all walks of life, independent of nationality, race, or sexuality. More than once, Disney has supported gay rights as opponents have threatened boycotts. The corporation held its ground on the subject so steadfastly that the boycotters eventually acknowledged failure. On social issues, The Walt Disney Company is left of MSNBC.

In spite of that, their financial donations during election years are decidedly one-sided. In 2012, the company directed 90 percent of their political buys to the Republican Party in Florida. Does this reflect a conservative point of view? The answer is actually no. Instead, it's simply a practical matter of good business.

The state of Florida has always enjoyed a unique working relationship with Disney. When the man himself purchased the land during the 1960s, he negotiated unique autonomy from the government. Despite Disney's control of its own lands, however, it's still susceptible to other changes in Florida. For example, much of the money Disney spent on lobbying Florida politicians went toward a specific goal. The company wanted to prevent the introduction of legalized gambling into the local economy. That would have changed the nature of greater Orlando from a family fun area to a more adult destination, ruining years of meticulous Disney strategizing. In other words, their lobbying during the 2012 election cycle was more about a single outcome than a point of view.

In this manner, the legacy of Walt Disney lives on. He too was a difficult man to pin down in a few words. He loved his employees almost obsessively, yet was stubbornly opposed to those who were socialist.

When forced to choose between finances and people, he avoided the issue as only he could, unveiling solutions that addressed both problems equally.

Disney deftly prevented the ascension of gambling from Native American-exclusive to readily available anywhere. By manipulating circumstances to protect Walt Disney World's interests, the company's leaders have followed Disney's blueprint for success. He believed in the laws of the land, but he also believed in adjusting the ones that made no sense. Disney did what he wanted with his land in Anaheim, California, and later with his new swampland in Florida. This behavior is what forced the United States government into an awkward situation.

Company leaders such as his brother Roy O. Disney have tried to honor Walt's legacy whenever possible. That's why the incongruity of EPCOT Center is so critical to this discussion. Before he died, Walt Disney envisioned a utopia, a place where people could come from anywhere on the planet and discover a new home. He bought almost 50 square miles of land in the greater Orlando area with the express intent of fulfilling this vision.

The politics of the suggestion were amazing then and remain so to this day. Walt wanted to build a community where everyone contributed to the greater good... which sounds a lot like socialism. There was a catch, though. Every member of the Experimental Prototype Community of Tomorrow would work for Disney. The unemployment rate would be zero. And Disney would involve itself in every part of the lives of its citizens. Effectively, it was the polar opposite of socialism: a uniform capitalism. Critics would even throw in some fascism when describing the terms. So, even Walt's utopia proved perplexing to the public.

The one thing Disney made clear is that he wouldn't be dealing with Russia. As noted during his House Un-American Activities Committee testimony, acrimony existed between an entire country and a single man, Disney himself. As he noted,

> I think they ran a lot of them [Disney films] in Russia, but then turned them back to us and said they didn't want them, they didn't suit their purposes.

In other words, communist countries hated Walt Disney's strident belief in capitalism almost as much as he hated their socialist economic system. When Disney announced plans for what would become EPCOT Center, he described a permanent World's Fair, which was a sound business strategy given Disney's dominance at the 1964 New York's World Fair. Card Walker, former chief operating officer at Disney, once stated "the nations of the world may participate on a permanent basis to demonstrate their culture and their products."

The above was true of everyone…except for Russia. Walt Disney was a man who could hold a grudge, and he was a patriot. Those two strikes were enough to exclude Russia from any potential EPCOT plans, even as China merited at least a modicum of consideration.

This was an interesting exclusion that Disney itself avoided in the wake of their founder's death. In 1976, a marketing executive famously noted that people loved the idea of EPCOT Center "all the way from Russia to the Philippines." This man, Jack Lindquist, would later become president of Disneyland, so he was in a position to know.

The statement signified a dramatic departure from Walt Disney's Red Scare mentality, even though it occurred during the height of the Cold War in the mid-1970s. By this point, Disney had adapted their plans to reflect a more international corporate philosophy. Even 40 years ago, sales revenue revealed the global popularity of the Walt Disney brand. For a permanent World's Fair to succeed, they'd need to develop a policy of inclusion, which was part of the early impetus for the social policies discussed above that are in place today.

The adapted initial plans for EPCOT Center expected 10 countries to participate at park launch. Disney expected to cull the list down from the 31 countries they'd initially pitched to have a presence at the second gate. Over time, park planners believed that the permanent World's Fair would expand to 30 countries, all of whom would share a part of their history and customs with the good people of Orlando, Florida.

This shift away from divisive politics toward an era of prosperity predicated on tolerance and inclusion sounds true in spirit to the original Walt Disney vision for the Florida Project. It sounds wonderful in theory and we now know after almost 35 years in operation that it works in execution as well. There was just one little problem, however, and you've probably already figured it out.

Have you seen *The Departed*, the 2006 movie directed by Martin Scorsese that stars Matt Damon and Leonardo DiCaprio? Since it earned a lot of money and won Best Picture at the 79th Academy Awards, the answer is likely yes. Just in case you haven't, the central conceit of the film is that the mob and the police are always fighting.

Since both organizations are wary of spies, each one tries a new kind of infiltration. They train someone impossibly young to join the other side too early to seem like a spy. Since each group has the same idea and the two young spies know one another, they cancel each other out, leading to chaos, bloodshed, and a confused audience wondering aloud when Mark Wahlberg learned to act.

The whole conceit sounds ludicrous, but it's actually based in fact. One of the best ways for moles to become spies surreptitiously is by joining at such a young age that they avoid suspicion. That's because their integration seems fluid and organic rather than forced. Does the premise sound comical? Of course. Still, if you're professionally obligated to worry about everything, this sadly counts.

During the construction phase of Walt Disney World's second gate, Disney hired 10,000 builders. The government had no cause to worry about any particular individual, yet they cast a wary eye on the entirety of the company's hiring practices for the Florida Project.

From the perspective of a government agent, Magic Kingdom attracted 13.2 million guests in 1981. While there was a downturn to 12.6 million in 1982, that's still a massive amount of people visiting the southernmost state on the East Coast. The introduction of EPCOT Center could theoretically double attendance for the region, which it did. In 1983, 22.7 million people visited Walt Disney World, and that number increased to 23.9 million by 1986. The second gate gave tourists twice as many reasons to visit Central Florida.

What's scary about that notion? The distance from Miami to Orlando is 235 miles. The distance from Miami to Cuba is 330 miles. EPCOT Center opened only 21 years after the Bay of Pigs conflict. That's roughly the same time gap from the release of the original *Toy Story* until today. Sure, it's a great deal of time, but for something as unforgettable as the United States and Russia reaching the brink of nuclear war, it's not a long enough time to forget.

From the FBI's point of view, about 10 million more people were ready to visit a place where 10,000 construction workers were breaking ground on a revolutionary project, EPCOT Center. The government had no way to block people from going, and they had no way of knowing the intentions of any potential guests. Also, the location Walt Disney selected in Florida also happened to be one of the closest major cities to a communist country. And that wasn't even the worst part.

In culling the list short list of countries for a permanent World's Fair, Disney executives quickly realized the obvious. After Greece and Africa, China provides the oldest, most fascinating civilization on the planet. They simply couldn't build a structure like EPCOT Center then exclude China. It would render irrelevant everything they had done and were hoping to accomplish.

This decision was like waving a red flag in front of a bull to government law enforcement officials. Now, EPCOT Center wasn't simply a security risk due to the population. A potential China Pavilion was also going to be authentic in tone, highlighting the ideals and beliefs of a people whose very way of life seemed decidedly un-American to the CIA and FBI in the late 1970s. What Disney proposed was the equivalent of not only embracing an enemy thousands of miles away but also inviting many of them into the United States as permanent guests. Sure, it's the alarmist evaluation of the matter, but FBI and CIA employees are paid to do exactly that.

Beginning in 1979, the federal government took such an interest in Walt Disney World that they worried Disney employees might notice. Despite this concern, the local Tampa Bay field office for the FBI requested more manpower and authorized man-hours.

Initially, the largest concern was that Disney would select Russia as one of the first countries represented at EPCOT Center. After Disney decided to honor their founder's wishes by resisting this possibility, the FBI saw a different shadow. They carefully studied Disney's plans for the various international pavilions. Something jumped out at them. Actual residents of foreign countries would become de facto citizens of the United States.

Disney planned to host a series of student ambassadors in the various housing facilities on site. In other words, the communists would have the ability to spread their dogma to people from other countries. Disney named their plan to have natives of each nation work at the pavilions as the World Showcase Fellowship Program. This innocuous idea that was mostly about authenticity scared a lot of people.

The particular anxiety was that the Chinese who were invited guests of America would spend a lot of time attempting to indoctrinate citizens. The FBI now perceived Walt Disney World, the Most Magical Place on Earth, as one of the most dangerous security breaches in the continental United States.

How do shadowy law enforcement officials handle surveillance for a locale as beloved as Walt Disney World? The answer is secretively. In fact, the information didn't become publicly available until the end of 2015, and it only occurred due to the dedication of a gentleman named Jason Smathers at MuckRock.com. On March 16, 2012, he submitted a Freedom of Information Act (FoIA) request for details about Disney records on file with the government. A bureaucratic nightmare ensued.

For three years, Smathers checked with the FBI. He posted the same FoIA request time and again, passed around to at least three agents during this timeframe. I presume that he persisted for a simple reason. He had to realize that their evasiveness indicated that he'd somehow struck gold with his request. On December 3, 2015, Smathers got to say something most people only dream about. He fought the law, but the law didn't win. The FBI was in the process of redacting the appropriate information from the secured documents prior to sending them to him.

Here are the details of the WSFP, the sponsorship system that terrified the FBI. Countries would have the right to pay $15,000 each (the equivalent of $52,000 today) to embark on a sort of foreign exchange program. Natives of these lands would work at EPCOT Center for 40 hours each week; 75 percent of that time on the clock would involve the actual job of being a Pavilion employee, and the other 25 percent would involve job training and local education involving currency, social interactions, and tourist behavior.

If this sounds innocent, that's because it is. Your local high school probably had a similar program, although I doubt they got $15,000 per exchange student. That's the high cost of working for Disney. The FBI's fear circles back to *The Departed*. They expected an influx of 200 to 250 people, although they used some shaky math to determine each pavilion would include about 10 residents of the applicable foreign nations. That's 20 to 25 each for the 10 pavilions, which turned out to be too high since only nine pavilions opened when EPCOT debuted, but the point here is about early indoctrination.

These students would come to America, learn our ways, and integrate into society. At that point, the FBI worried they'd be difficult to track and almost impossible to differentiate from ordinary citizens. Remember that computer technology in the 1970s was somewhere between Pong and Atari. A Vic-20 stood out as cutting edge tech for 1980. When EPCOT Center opened in 1981, people who worked there would gain all the appropriate identification to pass for Americans.

If there's one thing people in the espionage community hate, it's potential sleeper agents. The Tampa office of the FBI requested back-up. Their three requests were equal parts practical and panicky. They expected immediate notification if FBI's central headquarters learned of terrorist cells operating in the United States who might find allies at Walt Disney World. Tampa's law enforcement officials also asked for precise instructions about who would handle their processes in the event of emergency situations unfolding. Basically, if Chinese spies attacked anybody, the central Florida office wanted to know where they should report and who would be their handler. And they wanted a cryptonym (aka a code name) for any and all spy activities performed by potential insurrectionists at EPCOT Center.

While the FBI as a whole worried the most about *The Departed* scenario, Tampa's local office prioritized something different. The six foreign nationals running the show at each pavilion were a more direct threat in the short term than youthful sleeper agents. They wanted the ability to surveil these foreign nationals lodging within the walls of the Walt Disney World employee village. Their request never came to fruition since the international delegates in question preferred not to live like co-eds. So, they acquired residence in other parts of Orlando, like your average grown-up would do. This should've tipped off the FBI that people from foreign countries viewed their Walt Disney World assignments as dream jobs rather than a reason to force their political causes on strangers.

Also—and this is the funny part—many of the delegates selected to work at the various pavilions weren't even living in their home countries at the time. For the sake of convenience and/or pragmatism, many countries including China selected people living in America for their Walt Disney World pavilions. Such people offered a better understanding of the cultures of both their homeland and their current nation.

In fact, the most important delegate on the list was a restaurant owner in New York City. K.W. Poon became the Chinese representative at their EPCOT Center pavilion. The FBI carefully detailed as much of his bio as they could. The worst criticism they had of him on a personal level was the following. "Poon's primary interest in the whole operation was to make money." Yes, the dreaded

Chinese insurgent who the FBI feared would infiltrate our country and foment rebellion from within...was a huge capitalist.

Also, and this feels like pettiness from a bored FBI agent, Poon's file notes that his food wasn't very good. Also, his business was once shut down for a week due to its unclean food preparation practices. So, if you're ever enjoying a meal at Nine Dragons or Lotus Blossom Café, you're clearly not eating one of Mr. Poon's recipes.

Summarizing this entire series of events seems ludicrous on its face. The FBI developed a fear of foreign nationals that bled into every facet of society. During the late 1970s and early 1980s, the World's Fair concept wreaked havoc with the government's self-perceived concept of security. In addition to the issues at EPCOT Center, similar issues persisted in Knoxville, Tennessee, home to the 1982 World Fair. Perhaps the only people who understood and sympathized with the Tampa office were the FBI officials in the Knoxville, Tennessee branch. They too faced numerous inquiries and suspicions involving a Chinese delegate at their upcoming event. The two offices communicated frequently, hoping that the sharing of information would lead to better security measures and processes.

How well did all of this work? Well, we're not reciting a pledge to Xi Jinping every morning, so that's something. In reality, the American government wasted millions of dollars worrying about the Red Scare. Members of the FBI trained to prepare for situations that never had a chance of transpiring. China never had any intention of invading the United States, starting through Orlando, Florida. They didn't even care about sticking a Chinese citizen at the Chinese pavilion.

When push came to shove, China picked the most American option possible, a money-hungry New Yorker.

If the entire affair had been a James Bond film, 007 would have followed a dude for 100 minutes, only to realize it was an old school chum of his. Then, they'd have a shaken martini and talk about the weather. The FBI has done so much to protect our country that we could never thank them enough for their service. What happened with Walt Disney World, however, goes on their blooper reel.

CHAPTER FOUR
The Living Seas

Anything you can do, I can do better. At least that's how the saying goes.

The Walt Disney Company enjoys its status as the originator that all its imitators try to duplicate. That's how DreamWorks Animation wound up making a movie called *Antz*, an idea that Jeffrey Katzenberg had heard during his time as head of Disney's film division. It also explains why so much of the design of the Wizarding World of Harry Potter is familiar to Disney fanatics who remember potential park expansions from days gone by. They know that elements such as the dragon at Diagon Alley were originally part of expansion plans for Disney's Animal Kingdom. And SeaWorld itself? You'll see.

The unexpected truth here is that Disney is susceptible to such behavior, too. Just because they're generally the ones getting ripped off (er, imitated) doesn't mean they aren't above borrowing an idea here or there. Disney park planners have confidence that even when their ideas share surface similarities with those of competing parks, their theme will differentiate the product.

This justified confidence can get them into trouble. Such is the case with one of their most ambitious endeavors. Disney Imagineers looked on with longing as they watched SeaWorld become one of the favorite vacation destinations for families with small children. The Big Ideas folks at Disney dreamt of crafting their own attractions, ones that could sustain entire habitats for aquatic life. At the same time, the bean counters at the Mouse House sought to take back those vacation dollars the company had ceded to the first SeaWorld.

Back when SeaWorld first opened in 1964, Disney failed to anticipate the popularity of this concept with kids. Less than 20 years later, the opening of EPCOT Center provided them a rare opportunity to correct that miscalculation. Disney could build its own educational entertainment facility about underwater living.

In the process, they'd honor their company founder, Walt Disney, in two different ways. EPCOT Center itself was the living legacy they were building to honor his lifetime of great works and grander ambitions. And Walt always felt fascination with the water. That's why Submarine Voyage anchored the first massive expansion of Disneyland in 1959. Amusingly, executives from Anheuser-Busch at least partially gained their inspiration for SeaWorld from Disney's first water-based attraction. Almost two decades later, the park that was once an imitator had become a source of fascination to Disney employees in several different fields. They all agreed that an aquatic facility at EPCOT Center was a good idea.

Whether all these cast members were right or not is up for debate.

In the 25 years that have followed the original pitch, the facility we now know as The Seas with Nemo & Friends has become undeniably popular. It's also suffered through an identity crisis, a sharp decline in popularity, a return to glory thanks to clever theming, and several unfortunate incidents involving the animals who reside at the facility. While still an anchor piece of Epcot, its grand ambitions and astonishing architectural triumph have quietly taken a backseat to many of the issues that SeaWorld has faced. How did Disney build such an amazing facility, why did it struggle, and what does the future hold for Epcot's oddest pavilion? Let's dive into the complex history of what we once knew as The Living Seas.

If imitation is the most sincere form of flattery, then SeaWorld was the most popular flatterer in the amusement park industry for decades. Once onlookers in the entertainment industry understood the awesome drawing power of Disneyland, they embarked on a quest to build their own version of the premise. For George Millay, the founder of SeaWorld, and his staff, that led to the blueprints for an underwater restaurant.

These businessmen understood that Disney was on to something, and they viewed Disneyland as a kind of proof of concept for children's attractions of all sorts. Their plan for an underwater restaurant no longer seemed ambitious enough. Instead, they evolved the idea into an entire marine zoo in San Diego, California, only 93 miles away from Disneyland. Walt and his Imagineers felt a bit territorial about this turn of events, but they also admired both the temerity of the idea and its immediate popularity.

Kids really love cuddly fish, and that assessment was just as accurate in 1964, the year SeaWorld San Diego opened to the public. The park received 400,000 guests during its first year. It wasn't quite the 5.9 million guests who visited the Happiest Place on Earth that year. It wasn't even the 1.2 million who went to Disneyland during its first year, but a third of that total for what was effectively a fish zoo was quite remarkable.

From Disney's perspective, SeaWorld demonstrated proof of concept twice in quick succession. A decade after the San Diego park excelled, SeaWorld Orlando debuted. It popped up 26 months after Walt Disney World and was only five miles away. By now, Disney officials understood that one of the underlying strategies of SeaWorld was to place their parks close enough to Disney parks that vacationers would consider visiting both during the same trip. Disney execs were now doubly territorial. Of course, they had bigger fish to fry, so to speak.

After Walt Disney's death in 1966, his employees faced a seemingly impossible task. They had to honor his vision for the Experimental Prototype Community of Tomorrow without his direct input. All they had were his blueprints and the introductory video that he'd made about the project. Those weren't even the greatest challenges, though. They also had to figure out a way to pay for all of it.

Have you ever wondered why the Magic Kingdom came first at Walt Disney World? Well, it wasn't a magical decision that led to the Most Magical Place on Earth; it was a pragmatic one. Disney officials, including Walt Disney's brother Roy O. Disney, understood that they could never finance even a fraction of the company founder's ideas without a steady revenue stream.

The utopia that Walt imagined couldn't become any sort of reality until Disney could pay the bills. That wasn't an easy proposition at the time, either. For all its triumphs and name recognition, the corporation faced a great deal of cynicism as outsider observers loudly wondered about the future. Could The Walt Disney Company survive and thrive without Walt Disney? With the benefit of hindsight, that question seems laughable, but it was very real to Disney's staff during the late 1960s. That's also when they plotted the specifics of the Florida Project.

Once Walt Disney World opened to a rapturous reception from the East Coast residents who'd always wanted a Disneyland of their own, Disney execs had their revenue stream. And although the process took another 11 years, they would eventually take their best shot at honoring their founder's dreams for a better tomorrow. EPCOT Center wasn't quite the utopia Walt had promised, but it did combine two of his most important themes. It celebrated the various cultures of the world, and it attempted to educate visitors on the impending technologies that could change their lives.

The park we now know as Epcot blended education with entertainment, and the employees in charge of strategizing its future knew one thing for certain. SeaWorld officials had stolen a couple of key ideas from them over the years. The second gate at Walt Disney World would allow them to take some of those concepts back and re-establish them as Disney themes. They could build their own aquatic zoo, and it'd blow the competition out of the water. Hey, everything sounds good in theory. How an idea functions in real life is much harder to anticipate.

EPCOT Center opened exactly 11 years to the day after Magic Kingdom. The Future World portion that comprised the front half of the park focused on attractions that could simultaneously teach and entertain. The Living Seas Pavilion wasn't one of the initial facilities. It wouldn't debut for another five years, but the underlying structure was identical. That's because Disney execs had known since the late 1970s that they'd build their own SeaWorld-esque pavilion. As always, the issues of designing and paying for it delayed the proceedings.

The influences for The Living Seas were myriad. Disney looked at its own 20,000 Leagues under the Sea attraction for inspiration. They'd ported the attraction over from Disneyland to Walt Disney World in time for its 1971 debut, less than two weeks after the park itself opened to the public. Disney analysts believed that the fantastical elements of the Jules Verne story appealed to children. While park planners intended EPCOT Center to cater to adults, they understood that many guests would have kids to entertain. A fun version of their latest pavilion would make the day(s) spent at EPCOT Center feel less like school.

Since theming is everything at The Walt Disney Company, the intended pavilion was to embrace the heritage of aquatic mythology. Poseidon, ruler of the high seas, would greet guests at his underwater home. That visual would stand as the first thrill of The Living Seas. EPCOT Center visitors would see something akin to Atlantis. Imagineers drew illustrations for the entire showcase exhibit of the pavilion as an underwater domed city. Tubes would handle all the transportation in and out of the facility, and guests would enter and ride in giant water bubbles to reach the main area. If you're struggling to envision this and are a fan of classic British television, it's akin to the Rover from *The Prisoner*, something lampooned on *The Simpsons* as "The Computer Wore Menace Shoes."

From a presentation perspective, everything about this concept is sublime. Humans are fascinated by domes, and Walt Disney had at least considered placing the entirety of Epcot in a dome during its planning phase. The presence of this stunning underwater city would have dazzled all observers, even the ones who were long since jaded by most theme park attractions.

Similarly, the Poseidon theme was a fantastic way to reinforce the underlying premise of The Living Seas while introducing a grandiose element right at the start. The thunderous voice of a Greek God would have grabbed everyone's attention, making his informative speech of the importance of aquatic life unforgettable. It was the kind of showmanship and inventiveness that would have made Walt Disney proud.

Building the underwater domed city fell squarely under the header of "absolutely impossible!"

Don't get me wrong. Disney Imagineers tried. From the mid-1970s on, they knew that a water pavilion was going to comprise an integral portion of EPCOT Center. They crafted countless blueprints and illustrations of potential ways to bring the idea into reality. Unfortunately, working with water is one of the most difficult tasks on the planet. You may take indoor plumbing for granted, but it's true. The pipes that run into your home are an engineering marvel, and the difficulty in re-creating them is why so many parts of the world lack clean water.

The financial aspect is also problematic. Have you ever heard the story of *Waterworld*, the 1995 Kevin Costner movie? With a production expense of $172 million, it was the most expensive film ever made up until that time, but Universal Pictures hadn't budgeted anywhere near that much for it. They expected a cost of $100 million. The rest stemmed from the production staff's inability to maintain a steady supply of water on the set. Had they been familiar with the history of The Living Seas, they would have known about this issue in advance.

When Disney opened The Living Seas in 1986, they acknowledged that they'd spent $90 million bringing it to life. That's the equivalent of a little over $200 million today. It's also a lot more than they'd expected when evaluating the project. This was a recurring theme with EPCOT Center. Disney had projected that the entire park would cost $600 million. Even after they lopped off several pavilions from the opening year, it still cost an estimated $1.4 billion, more than double expectations. Still, the financing for The Living Seas was unprecedented. An opening day article from the *Orlando Sentinel* described it as "the single most costly project ever built in a Disney theme park."

On top of the shocking financial outlay, Disney had sacrificed virtually all of their grand ambitions for their SeaWorld-killer project. A domed underwater city was one of the first ideas to go. That's a given since it's a concept that's difficult to achieve today, much less in the early 1980s. They also discarded the transparent water vehicles, instead settling on the Hydrolator and Seacab, the former of which was a renamed elevator while the latter was just a themed Omnimover that would introduce guests to the oceanic environment.

Even the boisterous introductory speech from Poseidon fell by the wayside. Disney replaced it with a much more generic two-and-a-half minute video that felt like a PBS special rather than a booming Brian Blessed monologue. In three simple steps, they reduced The Living Seas from glorious celebration of mythic Atlantis with some *Sealab 2020* thrown in for good measure. Left in its wake was a pavilion that only parents could love while their kids wondered why they were doing this instead of riding Space Mountain.

The worst part is that like the Haunted Mansion before it, Disney had advertised the spectacular version of their underwater pavilion years before its arrival. They bragged of attractions such as the World of the Sea and Sea Base Alpha, the latter of which they declared would be a "futuristic undersea research station." Disney habitually under-promises and over-delivers because that's good business. With The Living Seas, they did the opposite, and it led to shattered dreams from park guests and park planners alike.

What went wrong with The Living Seas? Suffice to say that the situation unraveled quickly. The announcement above was featured in an advertising booklet from 1980, two years prior to the opening of EPCOT Center. By 1982, they'd already scaled back all their most quixotic plans for The Living Seas.

At this point, the construction team had explained in detail the challenges of building the underwater city originally promised. Presumably, these conversations mirrored ones that SeaWorld planners had while discussing their discarded underwater restaurant premise.

Imagineers leveled with their bosses about the impossibility of these (wonderful) ideas. Not coincidentally, the projected opening of this pavilion moved back a year, from 1983 to 1984. And they were still two years too optimistic. Working with water always comes with a few unexpected tribulations. More importantly, it comes with a cost.

The bean counters at Walt Disney World grew nervous as they started checking the cost itemizations for The Living Seas. They noted that one bill stood apart from the rest. In other to not just compete with SeaWorld but blow them out of the water, so to speak, Disney Imagineers intended to build a massive water tank. I mean a historically unprecedented water tank. And you know the one I mean since it's the one part of The Living Sea blueprints that became a reality.

Estimates for the water tank that would become the heart of the pavilion were stratospherically high. To pay for this single feature, the crux of The Living Seas, Disney would have to spend more money than they'd ever spent on anything before. Anything extravagant beyond the water tank quickly moved into the "absolutely not" category. Disney did get their money's worth on this particular item, though.

When The Living Seas finally debuted in 1986, then-CEO Michael Eisner trumpeted the massive aquarium that his company had built. Eisner stood in the cozy confines of a restaurant you've probably visited at some point, the Coral Reef. As he looked into the 22,000-cubic meter mother of all fish tanks, he bragged that the 5.7 million gallons included within aptly mimicked a Caribbean reef.

To punctuate Eisner's point about the majesty of the achievement, Disney president Frank Wells, who drew the short stick on this assignment, swam into view. He was in full wetsuit, and he carried scissors that he and a cast member in a Mickey Mouse wetsuit used to cut the ceremonial opening day ribbon. That tank already held 4,000 different fish, and Disney firmly stated that the total would double almost immediately.

The tank at The Living Seas was something that the competition had never managed despite the fact that they worked with water and marine life every day. Eisner and his team were right to take pride in their achievement. This new aquarium would remain the world's largest saltwater tank for almost 20 years until the Georgia Aquarium claimed that title in 2005.

The company also said something that remains true to this day. They built the underwater tank simply for the sake of learning. Disney hired experts in the field of marine life studies, and they constructed state-of-the-art research centers where the cast members watch and interact with the fish to this day. The company's employees take such pride in their giant aquarium that a version of it plays a central part in Disney's 2016 animated classic, *Finding Dory*.

Whether the scientific advances of The Living Seas make up for the overly optimistic initial claims is in the eye of the beholder. What's undeniable is that once the cost of the aquarium became unwieldy, Disney dismissed the other Big Ideas. Because of these concessions, their attempt to beat SeaWorld at their own game fell by the wayside, although The Living Seas instantly did become the sixth largest ocean manufactured by mankind. It wasn't as big as actual oceans, of course, but it was one of the most impressive feats ever accomplished by human construction to that point. The fact that Epcot still uses the body of it 30 years later reinforces Disney's greatness in building this phenomenal architectural triumph.

Even though sacrifices were made that lessened the pavilion from its potential status as the greatest thing in theme park history, it has stood the test of time. The Living Seas is something SeaWorld execs could only dream of bringing into reality. On opening day, guests witnessed cast members swimming through the depths of this artificial ocean. The Disney employees loved to display the nautical acrobatics possible within the enormous saltwater tank. With a wetsuit, an oxygen tube, and goggles, they could mimic virtually anything that a deep sea diver does during their time underwater.

The fun and games aspect of The Living Seas was a side benefit rather than a true purpose, though. Disney leaned on the side of science and education when they finalized construction details. The heart of the building, the portion that guests still enjoy today, is Sea Base Alpha. Imagineers crafted a quasi-futuristic "underwater" laboratory. It was so plausibly science fiction in look and style that a 1990s series called *SeaQuest DSV* actually filmed there. If it's good enough for Roy Scheider, it's good enough for me.

Sea Base Alpha offered several attractions, almost all of which were informational. The dive tank chamber and wave tank showed off the latest in underwater technology, highlighting the gear that would allow people to spend more time under the sea, and in a more comfortable manner to boot. Cast members would swim in these waters, demonstrating near-future wetsuits and the like. The wave tank also revealed the impact that the tide has on the otherwise unseen depths of the ocean floor.

The first floor of Sea Lab Alpha also included several aquatic modules. In these sections, guests could watch marine life in its natural habitat (the same was true of the marine life with guests). Ecosystems included a coral lagoon with countless starfish and other sea creatures, a viewing center for manatee interactions, some maps and educational videos/imagery, and an audio-animatronic whose sole purpose was to hype the future explorative value of underwater robots. Cast members also schooled observers in the nature of various fish.

The second floor was more of the same, sometimes literally. A couple of the main exhibits were (more than) two stories tall, so the higher floor provided a different view of the same activities. The Observation Deck was the seminal part of the top floor of Sea Lab Alpha. Its sole purpose was to display as many of the elements of The Living Seas aquarium as possible.

The design of Sea Lab Alpha wasn't happenstance. A few years prior, park planners felt pressured to throw out the grand design of a single immersive attraction. They'd intended to build an entire exhibition. It would begin the instant the park visitor boarded the transparent bubble and rode to the underwater laboratory. Their vision was of an all-encompassing underwater dome city experience. When the bean counters said no, many of the ideas Imagineers saved became truncated. Disney compartmentalized each one in a single section of The Living Seas. By definition, the modules made a trip through the pavilion modular in nature. The compromises lessened the impact of a visit to a water realm beneath Earth's surface.

While kids loved seeing the fish and remembered the Observation Tower and other exhibits vividly years later—and I speak from experience here—they bored quickly of the stuffier elements of The Living Seas. Watching fish is fun for a few minutes. Then, it gets old fast. As for the educational elements, Disney deserves tremendous credit for their decision to prioritize knowledge over entertainment. It was the opposite of SeaWorld's strategy. Unfortunately, kids have a heavy favorite in this race, and it's not knowledge. That's why Epcot's decision-makers eventually redesigned The Living Seas with a more child-friendly theme. We'll get to that in just a moment. Before then, let's discuss the pink elephant tap dancing in the living room.

"We won't commercialize sea life. There are no tricks here."

The above quote comes from a Disney spokesperson on the day that The Living Seas debuted. The employee was understandably proud of the seventh pavilion at Epcot, and they didn't back away from the pointed comments about other, more social water-based entertainment facilities. In fact, the PR rep quickly doubled down:

> Dolphins don't jump through hoops in the ocean. They won't do it here. Sharks don't go around eating boats or cities. They won't do it here. If you see some showgirl riding a porpoise, you're at the wrong place. Living Seas' focus is educational, not Hollywood.

Yes, that's a shot at SeaWorld, Universal Studios, and any number of low-rent amusement parks all in a single, vaguely condescending quote. Disney employees felt confident that their company was going to do everything right, and they felt this way due to past experience. A decades-long track record of excellence builds faith. But some things in life are beyond the control of even the greatest professionals.

I won't recount the Blackfish incident and how it's destroyed SeaWorld. Let's simply acknowledge that many individuals feel conflicted about the capture and harboring of mammals in small environments. Even as Disney sought to differentiate itself from SeaWorld, they employed some of the same practices, largely because they were commonly accepted in the mid-1980s.

Veterinarian Jay Sweeney was the focal point of Florida dolphin captures during this era. He was a co-director at Dolphin Services International, and he proudly proclaimed that he'd taken 80 dolphins out of their natural habitats. He performed these captures on behalf of theme parks like SeaWorld, knowing that the industry needed dolphins. They were the friendliest and most popular creatures at the parks, and they could perform tricks to boot.

When Epcot analysts plotted strategies for The Living Seas, they knew that dolphins would attract huge crowds. They hired the services of Sweeney and other dolphin trainers in the region, understanding that "training" was frequently code for "capturing" in the wild. That's the first misfire from Disney park planners. They didn't anticipate that removing a cuddly dolphin from its natural habitat can turn the creature into a killer. In their defense, who expects dolphins to turn feral?

During the 1980s, Sweeney was one of only a handful of federally approved supervisors for dolphin collection. He was also quite good at his job. The veterinarian captured a total of six dolphins for Epcot, two of which were female, and he did so in a much quicker timeframe than Disney had projected.

As for his qualifications, Sweeney was entrusted with the dolphins' caretaking by the federal government. His professional reputation was above reproach. A 1991 *Los Angeles Times* article detailed his capturing philosophies:

> Sweeney—one of the world's leading marine-mammal veterinarians—generally uses conservative methods for catching dolphins. He's never captured more than two or three dolphins on a given day; he personally monitors the animals for signs of shock and supervises their adjustment to captivity. "If I don't like the way an animal is developing, I always err on the side of safety and release it," he says. "We have never killed an animal in a capture operation."

Disney didn't hide their intentions, either. To the contrary, the company planned a photo shoot for Sweeney's dolphin hunt. They also filmed the events for posterity's sake. In the build-up to the opening of The Living Seas, nothing the company or Sweeney did was considered unbecoming. Had they only caught five dolphins, the entire perception of their dolphins captures might have been perceived differently.

The problem dolphin was named Bob. He was part of a school of dolphins along with Christie and Katie, the females, and males named Geno, Toby, and Tyke. Over a period of three years, Bob killed three of his siblings. A fourth died of non-Bob-related causes, leaving only the killer dolphin and what I'm confident was a very nervous sibling who never turned his blowhole on Bob.

I want to stress that the problems were not Bob's fault. If you take any mammal out of its natural environment, you shouldn't experience surprise if the creature lashes out. In Bob's case, he became too aggressive with his fellow dolphins. That behavior directly caused two deaths in 1987, and then, according to the *Orlando Sentinel*, he "roughhoused 12-year-old Katie, worsening her lung condition and leading to her death, preliminary test results show."

Animal-rights activists were understandably appalled. Disney adapted a transparent posture, with a corporate vice-president acknowledging that The Living Seas had a "pretty cruddy" record with dolphins. Since Bob killed both the females, Disney's dream of breeding the dolphins died with Bob's multiple sororicides. They later transferred him to the National Aquarium in Baltimore, where he promptly contributed to the deaths of two other dolphins before dying himself.

Disney had the legal right to try to breed dolphins again. So, they acquired a trio of them from the Navy. Another government agency had marked them for release. Disney stepped into a bureaucratic nightmare, and their attempt failed yet again. The dolphins currently hosted at The Living Seas were all acquired elsewhere. Nobody questions the treatment of these mammals, but the potential for negative publicity always exists as long as Disney continues to host them. They largely avoided the backlash that destroyed SeaWorld during the Blackfish fiasco since they don't host creatures of that size. Also, consumers trust Disney more as a brand. The benefit of the doubt has helped them remain above the fray.

On a lighter note, dolphin caretaking isn't the only problem Disney faced in populating their aquarium. Their original announcement for The Living Seas celebrated the presence of sea lions. Quick, when's the last time you saw a sea lion there? Sure, they're a staple at SeaWorld, but they're persona non grata at Epcot. The reason for that is simple. They stink.

In the early days of The Living Seas, cast members quickly came to resent the presence of the sea lions. They seem so harmless and fun from a distance. Up close, they're loud, obnoxious, and odiferous. Over the years, Disney quietly phased out the sea lions part of the habitat, repurposing it as a home for manatees.

Today, the location hosts Lil Joe, widely considered the most famous manatee in aquatic history. For 25 years now, he's enjoyed his celebrity status at several zoos and theme parks. Disney hosts him at the pavilion to save the manatee from getting stuck in the pipes again.

Yes, Disney would rather have a clumsy, fat manatee who is terrible with directions instead of sea lions. That says everything about how awful they were as residents of The Living Seas. *Finding Dory*'s sea lions have a presence in the movie due to the popularity of fish tales involving the terrible behavior of these creatures. It's an open secret at Epcot that they quickly wore out their welcome. Take a moment to consider how awful they must have been. Even Bob received the benefit of the doubt for several years and a few murders.

As attendance dwindled at The Living Seas, cynics derided it with a new, commonly used nickname. What was once the most expensive building at Epcot became known as The Dead Seas due to its lack of traffic. Disney officials at first were part of the problem rather than the solution.

When Epcot lost the original sponsor for the pavilion, United Technologies, Disney's notoriously cheap CEO, Michael Eisner, chose to shut down one of the most expensive elements, the Seacabs. Why this decision made sense to anybody is up for debate. Without the introductory portion of The Living Seas, it felt incomplete, as if a person were catching only the second half of a movie they'd never seen before.

The haphazard deletion of an iconic element of the pavilion escalated its rate of descent. Anyone who had visited Epcot in the past resented the absence of the Seacabs. When new guests exited The Living Seas, they felt like they'd missed a key step. The entire premise had collapsed from the grandest of ambitions to the least popular building at Epcot. Park planners accepted that they needed to change the fundamental nature of the entire facility, albeit at an acceptable cost.

The central flaw with The Living Seas is that it didn't appeal to kids the way that Disney had intended. Rather than building a SeaWorld-killer, they'd compromised far too many times. All they had left was a building whose fish population outnumbered its human guests. They had to act to reinvigorate one of the costliest constructions in company history.

The solution came in the form of a clownfish.

The release of *Finding Nemo* presented an opportunity to Disney officials. While they wouldn't purchase Pixar until 2006, the two corporations enjoyed the type of friendly relationship to be expected of the two finest animation studios in the world. In many ways, Pixar was the second coming of the Disney Brothers Studios, the 1920s business that grew into The Walt Disney Company. They were story-focused and family-oriented. Both of those are hallmarks of the Disney brand, and that's why Epcot execs took notice during the summer of 2003 when *Finding Nemo* dominated the box office.

Themed attractions at Epcot weren't historically unprecedented at this point. *The Lion King* already had a presence at The Land. What park planners plotted was new, though. They strategized on the best way to draw attention back to The Living Seas, and they acknowledged that the most effective way was through theming. The pavilion would become The Seas with Nemo & Friends.

In order to make such a drastic change, Disney restored some old ideas with a new coat of paint. The cobwebbed Seacabs could return anew, albeit with some alterations. The updated Omnimovers would point to the sides rather than straight ahead, drawing eyes to the walls the way they should have all along. They'd also receive different theming. Out was the clinically scientific vibe of deep-sea exploration. In was the Clamobile, a Nemo-appropriate ride cart that was warmer and more inviting to younger guests.

Beyond the surface changes, the former Seacab delivery system underwent a dramatic modification. The "walls" of the aquarium disappeared, for the most part. The majestic views of the fish contained within the water world vanished. In their absence, scenes from *Finding Nemo* sprang to life. The exception is at the end of the ride, and even that combines with artificial imagery. Disney park planners were acknowledging the obvious.

SeaWorld's strategy of catering to children first and then their parents afterward is the correct one. Disney's always known that, of course, and it's standard operating procedure for most of their attractions. They tried to act highbrow with The Living Seas, and the strategy cost them.

What does the future hold for The Seas with Nemo & Friends? The important factor to remember is that the facility is still wildly expensive to operate. Filtering, cycling, and maintaining that much water on a daily basis is a nightmare. Imagineers are also stuck with the architectural choices of the early 1980s. More than 30 years later, technology has come a long way in this field. Park planners don't have the ability to rebuild the aqueducts, though. They're stuck with the plumbing they've got.

Short of tearing down the entire facility and starting from scratch, something they're extremely unlikely to do, Epcot strategists are left with what's on hand. That's unfortunate because they still own the blueprints to one of the most imaginative, potentially spectacular pavilions they've ever built. I'd go so far as to say that if they somehow could build the underwater dome they illustrated during the 1970s, it'd instantly become the most popular building at the entire park.

In its stead, all they can offer at The Seas with Nemo & Friends is a bunch of seagulls parroting, "Mine! Mine! Mine!" There's also a charming themed ride that would feel right at home at Magic Kingdom, but that makes its presence at Epcot that much odder. The themed version is a far cry from what was announced for the park and directly counters the Disney spokesperson on opening day of The Living Seas. That person claimed Disney would never commercialize sea life. Reconciling that with Nemo theming is equal parts frustrating and sad. Somewhere along the way, Disney tried to beat SeaWorld at its own game. They badly lost their way, and only a clownfish kept the most expensive pavilion at Epcot from being called The Dead Seas today.

CHAPTER FIVE
Hollywood Studios

When you think of the various Disney theme parks, distinct premises reflexively enter your mind. Disneyland is the place that started the craze, the proverbial Happiest Place on Earth. Magic Kingdom is the East Coast equivalent, the Most Magical Place on Earth as well as the most popular theme park on the planet. Disney California Adventure is where the majestic mountains of Radiator Springs provide the backdrop for Cars Land. Animal Kingdom is the place where Disney created a hybrid of zoo and amusement park. And Disneyland Paris is the "cultural Chernobyl" once known as EuroDisney, which isn't a good thing.

For better or for worse, most Disney gates enjoy a specific identity. One of the parks at Walt Disney World is the glaring exception. When theme park tourists think of Disney's Hollywood Studios, ambiguity and hazy memories are all that arise. Once, the Earffel Tower stood as the landmark identifying Disney-MGM Studios. Later, the Sorcerer's Hat lorded over the central hub, reminding guests where they were. Today, a hodgepodge of attractions, many of them quite good, comprises the core of Disney's third gate. But Disney's Hollywood Studios continues to lack definition. It is the rare Disney theme park that suffers from an identity crisis.

As everyone knows by now, park planners have Brobdingnagian ambitions for the third gate. It will eventually emerge from its current cocoon as a largely disappointing half-day park into a resplendent butterfly, ascending into greatness thanks to the impending presences of Star Wars Land and Toy Story Land, a pair of locations that sound like the most desirable spots in a game of Movie Monopoly.

In the interim, critics rightfully assail this gate for its incongruous nature. Ostensibly a movie lover's paradise, some of its most famous attractions center on a half-century old television show and a rock band whose musicians are all approaching their 70th birthday. Meanwhile, some of the attractions with direct ties to classic Hollywood have shut down in order to make way for potentially more popular events that move Hollywood Studios even further away from its core concept.

How did such a noble, grand idea for a theme park erode so dramatically over time? What led to the fall of Disney-MGM Studios? Why have not one but two different iconic landmarks become deconstructed? And will the park rise like a phoenix from the ashes once it receives the fresh infusion of entertainment from Pixar and Star Wars? Or is this another desperate ploy that is doomed to fail?

Simply stated, most Disney theme parks succeed to a degree beyond all rational expectations. Why hasn't Hollywood Studios? Let's investigate its origins to learn why Imagineers suffered a rare misstep in park introduction, one that was entirely beyond their control. We'll detail the messy public divorce between two major corporations that undid this new park before it ever began. Then, we'll look at the steps they've taken since 1989 to bandage the festering wound that is the half-day park at the third gate. Finally, we'll look forward to make some educated guesses about how much Star Wars Land and Toy Story Land will mean to the bottom line of the park. Here, then, is the shaky history of Disney's

Hollywood Studios, the most organic yet also most awkward of all the company's theme parks.

Competition brings out the best in people. This belief has become one of the linchpins of the business world to the point that the laws of the land expressly prohibit monopolies. Even The Walt Disney Company, which owns a functional monopoly in the theme park industry, is subject to outside pressure. Whether they should worry about what other parks are doing is up for debate.

The numbers reflect the dominance of Disney's Parks and Resorts division. In 2015, the company claimed 137.9 million in attendance despite the fact that many of their parks cost $100 or more to visit on a single day. Putting that number in perspective, the company that claimed the second largest traffic in 2015 was Merlin Entertainments Group, owners of the London Eye, the Orlando Eye, and Alton Towers Resort. Despite including much cheaper tickets for their attractions, Merlin Entertainments Group earned "only" 62.9 million in traffic in 2015. That's a difference of 75 million visits between first and second place in the industry, which means that the gap is larger than the amount of guests who actually attended all of Merlin Entertainments Group's venues.

The only important aspect of theme park tourism is that the gap between Disney and its closest rival is expanding rather than contracting. Disney traffic grew 3.6 million year over year whereas Merlin Entertainments Group was effectively flat at 100,000 tourists gained. That's why more financial analysts pay attention to the heralded company that finished in third place on this list.

Universal Studios Recreation Group, a division of NBCUniversal, has grown at an exponential rate ever since they acquired the license to the Harry Potter franchise. This fact is important for two reasons, one of which we'll discuss below. For now, what matters is that from 2009 to 2015, Disney theme park traffic increased 18.8 million. Universal Studios garnered 21.1 million new vacationers.

On the surface, the gap seems modest, but that's because I've only provided you with half the story. In 2009, Disney's Parks and Resorts division enjoyed attendance of 119.1 million. Universal Studios was only at 23.7 million. Ignoring percentage growth, which sometimes paints an inaccurate picture of a situation, Universal Studios has added more park guests than its more storied opponent in the oligopoly of theme park tourism. It managed this feat despite starting at a wildly disadvantageous position.

In short, Universal Studios is eating Disney's lunch in the game of expanding the brand. In a way, this situation is prophecy self-fulfilled for Disney execs. What's happening today is something that park planners prophesied a long time ago, all the way back during the mid-1980s.

In 1985, Universal Studios enjoyed widespread success in mainstream cinema. Their biggest hits of 1985 were timeless classics such as *Fletch*, *Out of Africa*, and *Back to the Future*. Meanwhile, Disney bet their summer box office campaign on...*The Black Cauldron*. It was so unsuccessful that the company's board questioned the viability of Disney animation moving forward.

Disney vs. Universal Studios is an ongoing battle in Hollywood. Fans of the latter institution note that their idea of a Hollywood amusement park dates back to 1915, the first time guests had the opportunity to tour the studio backlot. Disney fans (rightfully) note that Disneyland became the first true theme park in the world at its 1955 debut. Roughly a decade later, Universal Studios copied some of the concepts of theme park attractions, co-opting them for what we now know as Universal

Studios Hollywood. Since July 15, 1964, the two companies have warred against one another as the two most heralded businesses in the theme park industry.

Then again, the battle has been largely one-sided. Before Harry Potter, saying that Universal Studios was competing against Disney for theme park tourists is the equivalent of my saying I'm competing against LeBron James to become the greatest living basketball player. Even if it's true in the most technical sense, the only commonality between us is that we've both shot a ball into a hoop countless times. Just because Universal Studios had a clever Jaws encounter didn't make it the equivalent of the Happiest Place on Earth.

Despite the inequality between the two theme parks during the first two decades of existence for Universal Studios Hollywood, Disney execs kept a wary eye on their foes. When rumors arose of a new Universal theme park in central Florida, it was more-or-less expected. Similar rumors had proven true during the 1960s only 35 miles from Disneyland. Why wouldn't Universal Studios once again follow Disney's lead by building a theme park near Walt Disney World? Once again, Disney execs felt the surge of adrenaline that stems from competition. And it led them to explore a new avenue for a theme park, albeit one that returned the company to its roots.

Here's a bit of Academy Awards trivia for you. Edith Head has won the most Oscars of any female, garnering eight for her amazing designs. Amazingly, that's twice as many as Steven Spielberg, who has won three times among his 16 nominations. Alfred Hitchcock, my all-time favorite Hollywood personality, was nominated five times but never won. Meryl Streep is at three wins and counting after 19 nominations.

I say all of this to offer points of comparison for a trio of staggering facts. Walt Disney once won four Academy

Awards in a single year. That's more than either Streep or Spielberg has managed during their entire legendary careers. It's also a record for most wins in a single year. Amazingly, those victories amount to only a fraction of Walt's total Oscars. He Walt won 22 times out of 59 (!) nominations. During his most staggering run of achievement, he won an Academy Award during eight consecutive nominations, as much as Head won during her entire career.

Nobody in the history of Hollywood has the pedigree of Walt Disney.

That's what makes the variation at Disneyland so marvelous. Rather than tether all his themed lands to specific intellectual properties from the already massive Disney library in the early 1950s, he offered sections that had tremendous variety, and set the path for those who followed him as Disney park planners.

The Pirates of the Caribbean is a great example of an attraction for which Walt provided oversight but didn't live long enough to enjoy its debut. Almost 40 years later, it became one of Disney's most popular film franchises. Other cinematic crossovers involving the Country Bears and the Haunted Mansion, neither of which did well, plus upcoming projects such as Jungle Cruise and, well, the Haunted Mansion, reflect how much modern Disney employees embrace their company founder.

Walt plotted a course for his themed lands that didn't conform to any single blueprint. At times, he was almost stubbornly averse to tethering his parks to Hollywood convention. And that's odd in and of itself since so many of the early Disneyland attractions employed techniques Disney and his Imagineers learned by making films. From the founder's perspective, he'd already told stories in that sort of format. Disneyland and the planned Experimental Prototype City of Tomorrow afforded him new avenues to entertain children of all ages.

In the aftermath of Walt Disney's death, Imagineers tried to honor his legacy to the best of their ability. The result of their efforts was Walt Disney World. The place that their founder envisioned as a functional city evolved into something different initially before eventually becoming a version of his intention. First, a theme park originated there on land conceived as a targeted utopia. Afterward, part of the vision of EPCOT came to fruition at EPCOT Center's World Showcase, which did include a version of a permanent World's Fair.

In the time between Magic Kingdom's arrival in 1971 and EPCOT Center's start in 1982, most of Walt Disney's ideas had devolved. They became debased by people who didn't understand them quite like he did. After EPCOT Center opened, the big thinkers at Disney faced a single difficult question. "What's next?"

In choosing the next steps for Disney's theme parks, one thing they recognized is that Walt was the one who specifically chose to disassociate his parks from Hollywood. The park planners who followed him felt no such compunction. If anything, they lamented that they now had three parks operating on a daily basis, yet none of them included specific ties to the work of Walt Disney. Sure, there were tributes and derivatives everywhere in sight at Disneyland and Walt Disney World, but none of them screamed classic Hollywood.

With plenty of land to spare in Orlando, Florida, everyone at Disney agreed that their third gate should have more ties to the early days of the Disney empire. They built a checklist of items they could use to populate this new park. Virtually all of them would enjoy ties to the Hollywood of old as well as modern movies and television shows. Disney architects crafted a proposal for their next enterprise, a celebration of all things cinema.

Still, nobody was pushing for the third gate until rumors of a Universal Studios Florida grew louder in 1985. While the park itself wouldn't open to the public until 1990, corporate execs at Universal started evaluating land acquisitions by 1985. Disney still had employees familiar with this process from their own Florida real estate deals two decades prior. Through their local connections, they quickly uncovered the truth. Universal Studios Florida would become a real thing within a couple of years.

That timeline proved too ambitious for Universal. Still, Disney execs felt pressured to act. That's how competition can impact a business, even one that is light years beyond the competition in terms of market share. A recent comparison is Facebook's acquisition of Instagram, which performed services Facebook could already do but in a much more user-friendly way. Part of running a viable business is understanding how quickly competition can overtake you. Disney execs possessed a better feel for this practice 30 years ago than they do today.

Once Disney officials verified the mere possibility of a new theme park within 10 miles of Walt Disney World, they proactively moved to step on the neck of the competition. They already had the plan in place, too. Better yet, the ideas driving the third gate were the stuff of Disney royalty. I'm not referencing any princesses, but instead then-CEO Michael Eisner and Disney Imagineering executive Marty Sklar.

During the lead-up to EPCOT Center, Sklar and another Imagineer, Randy Bright, pitched several ideas of pavilions. One of them stood out from the rest. It was called "Great Moments at the Movies." You know (or rather knew it) as the Great Movie Ride. This "pavilion" would have created a direct connection between EPCOT Center and Disney's storied movie history. It also would have celebrated the other classic tales of the golden age of cinema. The only problem with the pitch was that it was too good.

Eisner and Sklar quickly agreed that Great Moments at the Movies would feel like a cheat as a pavilion. Instead, it was a concept worthy of an entirely new theme park. That was the plan that wound up in mothballs before Universal Studios unintentionally started a new theme park arms race when they acquired land in central Florida. At that point, Disney officials decided that they already had a premise that was construction-ready. All they needed to do was flesh out a few more ideas, and they could break ground on a third park.

Fate was even in their corner at the start of the new venture. In 1985, Disney entered into a new agreement with Metro-Goldwyn-Mayer Studios (MGM) that gave the Mouse House the rights to the MGM name. Disney could use the historic, world-famous MGM name and Leo the Lion logo for their new theme park. It was a savvy acquisition that opened up the celebrated body of the MGM library.

Disney so loved the idea that they planned to do it twice. They would open a second gate at Disneyland around the same time as the third gate at Walt Disney World. This location would sit in Burbank, roughly 35 miles away from Disneyland, and would be called Disney-MGM Studio Backlot, a clever shot across the bow of Universal Studios. As Universal attempted to attack Disney in Florida, the Mouse House would target them in the greater Los Angeles area. Theme park tourists on both coasts could enjoy Great Movie Rides.

Divorces are always hardest on the children. In the case of The Walt Disney Company, their marriage with MGM barely lasted long enough to enjoy the honeymoon. After consummating its union with Disney in 1985, MGM quickly met somebody else. The hussy was named Ted Turner, and he seduced MGM for a brief period in 1986.

The man colloquially known as the Mouth of the South had a single purpose in the MGM purchase. He coveted the movie catalogue that would fittingly become the cornerstone of Turner Classic Movies, whose presence has been integral to the existence of the Great Movie Ride since its inception. Turner purchased the company, acquiring tremendous debt in the process. In order to negate some of his debt, Turner flipped many of the assets from his MGM acquisition quickly. Within months of his becoming owner of the fabled corporation, he'd stripped it for parts.

Once and future MGM owner Kirk Kerkorian quickly purchased some of those pieces from his former boss. From them, he rebuilt an entertainment division known as MGM/UA Entertainment. It was one of those loathsome 1980s corporate deals that impacted the lives of thousands of employees while a few multi-millionaires found new avenues to billionaire status. What nobody at Disney could have anticipated is how much that transaction would impact their planned third gate.

Light on capital and seeing shadows everywhere, the new corporate overlords on the board of MGM/UA Entertainment hated the Disney deal. It heavily capitalized on their brand without offering much tangible benefit in return. Disney's licensing contract wasn't satisfactory in their estimation. They sought legal excuses to break the deal. In reality, they were perfectly willing to extend the contract as long as Disney paid them more money. Understandably, Walt's team wasn't interested in doing so. From their perspective, they'd signed a legally binding document that was mutually advantageous to both parties. In their view, Disney-MGM Studios and its California twin would anoint MGM as the carrier of the torch of classic Hollywood cinema.

At the time, MGM/UA was all about the Benjamins, though. They found ambiguous language in their contract that allowed them to sue Disney. The point of conflict was, humorously enough, the *Ernest Goes to...* movies.

From the very beginning, Disney intended their third gate to double as a functional movie and television studio. They would film productions there while allowing park guests to watch the process unfold. Sometimes, the proverbial live studio audience mentioned as legal jargon during the credits for sitcoms would be people visiting Disney-MGM Studios for the day. MGM had a problem with that, though it wasn't the breaking point.

The backlots at Disney's new third gate at Walt Disney World were ready well before the park itself opened to guests. Rather than waste unused space, Disney embraced the filmmaking process by scheduling productions at their new Florida studio. One of the most heralded ones at the time was—I'm not joking—*Ernest Saves Christmas*.

In 1987, comedian Jim Varney capitalized on the surging popularity of his too-stupid-to-live but well-intended character, Ernest. The first outing, *Ernest Goes to Camp*, was a box office hit that grossed over $23 million on its modest $3 million production cost. One of the things Disney loved about the Florida studio lots is how well they could control the costs of their Touchstone Pictures productions. Even as Varney's salary escalated due to delivering a critically praised box office hit (again, I swear this is all true), *Ernest Saves Christmas* cost Disney only $6 million to produce. It was another hit, grossing approximately $28 million.

This financial gain was a Pyrrhic victory. The MGM attorneys attacked the language of their signed contract with Disney. They maintained that their business partner didn't have the right to produce movies at the new theme park. An MGM spokesman argued, ""We thought we were licensing it to an attraction, a mock studio." Effectively, MGM sued Disney for $100 million in 1989 because they signed a partnership agreement that meant Disney-MGM Studios was supposed to be fake, not real. From their perspective, they wouldn't willingly enable a competitor in the film industry to create movies at a joint site without compensation.

Disney execs felt that the claims were laughable. They believed that they'd been forthcoming about their intentions from the beginning. Instead, they considered MGM's lawsuit as a money grab from a cash-poor corporation. More important, Disney was salty about the attempt to break the contract. They fought back in a vicious manner.

One of the central tenets of the contract signed between MGM and Disney was that the latter company could use the MGM name for theme parks. Once Disney learned of MGM's plan to sue to break the contract, they targeted their opponent's flank. It was an open secret in the tourism industry that MGM planned to build a 4,000-room resort in Las Vegas. Part of their new project would involve a fake studio tour, exactly the sort of thing they'd contractually agreed to let Disney build using the MGM name.

In the annals of corporate intrigue, this situation was one of the most cut-and-dried from a legal perspective. Disney had licensing rights for MGM, which meant that any attempt by the latter company to build anything with even surface-level similarity to Disney-MGM Studios would violate the terms of the agreement. The Vegas resort including a studio tour was a flimsy argument, but it was still an unmistakable contract violation. Disney sued in September of 1989, a month ahead of the $100 million countersuit from MGM that they'd anticipated for three years.

The matter wasn't settled for good until October of 1992. At that point, a judge ruled in Disney's favor. They could continue to use the MGM name and logo per the contract agreement. The judge also sided with MGM with regard to what we now know as the MGM Grand in Las Vegas. Disney couldn't block its opening, although

Eisner himself pointedly stated that his company might sue again if the studio tour became too much like anything from Disney-MGM Studios.

Legal experts agreed that the judge's ruling heavily favored Disney, and the outcome of the entire situation was that both corporations wasted a tremendous amount of money on legal fees for what was a clearly defined contractual issue. MGM simply wanted to get out from under the contract or get paid more from Disney. Ultimately, they failed on both counts.

The problem with any divorce is that it has a detrimental impact on the children. In the case of MGM vs. Disney, the moment MGM tried to break their ironclad contract with Disney, the die was cast. Disney-MGM Studios, ostensibly a joint venture between two of the most iconic movie studios ever, was revealed as another petty Hollywood squabble between power players. Disney's third gate suffered from the beginning due to this awkward schism in the relationship.

On May 1, 1989, Disney-MGM Studios opened. Within six months, the two principal studios involved were suing each other for an incomprehensible amount of money. You can imagine the turmoil this caused behind the scenes at the park. Many of the ideas involving MGM properties underwent obscene amounts of scrutiny to guarantee that they didn't provide MGM additional wiggle room in the legal battle. In the end, virtually all ties to MGM beyond the park name, logo, and the Great Movie Ride were stripped from the initial attractions. And there were only two of those.

Despite all the grand ambitions Eisner and Sklar had for their new park, the legal wrangling negated potential ideas. When Disney-MGM Studios debuted, theme park tourists enjoyed precisely two options. they could board the Great Movie Ride or they could take the Studio

Backlot Tour. For all the criticism the park we now know as Disney's Hollywood Studios receives today, its 1989 debut was dramatically worse. The studios involved were embroiled in a classic Hollywood feud that caused both parties to behave in self-destructive fashion.

The Great Movie Ride, which had once stood out as a grand idea for an EPCOT Center pavilion, instead became the signature attraction at Disney-MGM Studios. And even its presence was awkward. Many of the ride's high points celebrated MGM classics such as *The Wizard of Oz*. By the time the ride opened, Disney execs hated the thought of giving MGM that sort of credit for their library. Even worse, some of the early parts of the Great Movie Ride suffered from technical malfunctions that were difficult to correct. As unbelievable as this statement may sound, in the early days of Disney-MGM Studios, the greatest unsullied accomplishment was an *Ernest Goes to...* movie.

That's a bit of an exaggeration. Since Florida lacked a natural film presence at the time, Disney execs scrambled to put their new space to good use. They wound up rebooting *The Mickey Mouse Club* with new episodes filming at Disney-MGM Studios. They also offered park visitors the opportunity to watch various animated projects unfold. A few lucky folks can honestly say that they watched *Lilo & Stitch* during the illustration process.

The other odd type of television production during the early days of the third gate involved professional wrestling. Ted Turner, Disney's former partner for a few months in the endeavor, acquired World Championship Wrestling, too. His advisors asked if they could hold WCW television tapings at the park. Disney's newest development offered better television production values than southern arenas. Starting in 1993, WCW filmed several weeks' worth of television tapings. Since wrestling outcomes are determined in advance, guests attending these live shows would know upcoming wrestling champions well ahead of the rest of the fans.

Of course, wrestling icons such as Sting, Macho Man Randy Savage, and Hulk Hogan weren't the most famous celebrities working at Disney-MGM Studios. The aforementioned *Mickey Mouse Club* dominated the A-List, although nobody realized it at the time. Disney cast members from the reboot of the Walt Disney television show included Christina Aguilera, Ryan Gosling, Keri Russell, Britney Spears, and Justin Timberlake. In hindsight, that was the greatest achievement during the early days of the park.

Even the landmarks from Disney-MGM Studios and Disney's Hollywood Studios have suffered from the MGM curse. In 2016, park planners tore down the pointless Earffel Tower. The original landmark denoting the third gate was supposed to stand as its answer to Cinderella Castle at Magic Kingdom and Spaceship Earth at Epcot. That structure, the Earffel Tower, doesn't even exist at Walt Disney World today. The 130-foot tall construct was a replica of the original one at Walt Disney Studios in Burbank, California. And the water tower didn't even contain water!

The primary purpose of the Earffel Tower was to provide a distant visual of the park. It also functioned as a highlight of the Studio Backlot Tour. Once that attraction closed in 2014, the landmark seemed like a waste of space. Disney does still have a variant of the original Earffel Tower in operation at Disneyland Paris, though.

Part of the reason that the Earffel Tower diminished in importance was the arrival of the Sorcerer's Hat in 2001. The iconic accessory debuted in "The Sorcerer's Apprentice" segment of *Fantasia*. More than 60 years later, Disney constructed a 122-feet tall representation of it as the new "face" of Disney-MGM Studios. People hated the new landmark. I cannot stress this point enough. I always thought it was kind of cool, as I'm a huge fan of *Fantasia*. Over the years, I discovered that the quickest way to pick a fight in a Disney online community was to praise the hat.

Why a giant blue hat caused such animus among theme park tourists is up for speculation. What we know for certain is that Disney eventually ceded to the vocal majority on the topic. In 2015, 14 years after its arrival, a construction crew started to take the hat apart piece by piece. It's a sad turn of events for the second potential Disney landmark at the third Walt Disney World gate.

If you're scoring at home, Disney-MGM Studios started with two attractions in 1989. Over the years, they've had two primary landmarks, not counting the re-creation of Grauman's Chinese Theatre at the now-shuttered Great Movie Ride. Amusingly, the Great Movie Ride too had ties to Ted Turner. In the aftermath of Turner's MGM acquisition and resale, he controlled the rights to Warner Bros. films created prior to 1959 and MGM films released prior to May of 1986, the point of (second) sale for the corporation. Turner capitalized on his new media library by launching Turner Classic Movies (TCM), a staple of cable television since 1994.

In 2014, Turner's business bravado again led to a business relationship with the Disney company. TCM licensed new content for the Great Movie Ride, which had recently turned 25 years old. In exchange, TCM would receive endorsements throughout the pavilion as well as sponsorship credit for the ride. The late TCM host Robert Osborne lords over the proceedings as the narrator. By this strategy, Disney managed to reinvigorate the remaining original attraction from Disney-MGM Studios—until, of course, they decided to close it.

The bigger change occurred a few years prior, though. Even before the hat and the tower were taken apart, Disney park planners finally acknowledged the obvious.

The ugly mess caused by the infamous legal battle with MGM left the third gate in turmoil. Guests subconsciously associated it with failure and chaos. In an attempt to remove the stigma, Disney dropped MGM from the park's name in January of 2008. Even this proactive move failed to redeem the reputation of the park. Instead, a popular weekend gathering triggered the events that have placed Disney's Hollywood Studios on its current trajectory.

Disney employees always enjoyed a warm relationship with George Lucas, the creator of the Star Wars franchise. Coincidentally, this union began around the time that Disney Imagineers started to plan the next Walt Disney World gate during the mid-1980s. Lucas interacted with Imagineers on a concept that eventually evolved into Star Tours. It quickly became one of the most popular attractions at each Disney park that hosted a version.

One of those locations was Disney's Hollywood Studios. The first iteration of Star Tours began there in 1989 and lasted until September of 2010. At that point, its replacement was ready. That sequel, known as Star Tours: The Adventure Continues, stands as one of the most brilliant ride designs in motion-simulator history, because each ride is variable with more than 100 permutations possible as of late 2015.

Lucas loved the ride and relished his relationship with Disney. He believed that they were the one movie studio that loved and respected cinema the way that he did. In 2012, Lucas sold Lucasfilm to Disney for $4 billion. The rest is box office history as the release of *Star Wars: The Force Awakens* shattered virtually every major box office record in existence. The transaction also had a major impact on Disney's Parks and Resorts division.

Starting in 1997, the Mouse House held Star Wars weekends wherein they encouraged fans of the movie franchise to visit Disney's Hollywood Studios. During these events, actors from the movie would appear, and cast members would portray characters from what was then a trilogy. They'd also dress up in new costumes that placed iconic Disney characters in Jedi outfits. It was a cosplayer's dream.

Since the third gate rarely enjoyed huge attendance numbers relative to Epcot and Magic Kingdom, it was the perfect location to host these weekends. By 2003, Disney understood that they had a hit on their hands. Star Wars Weekends were spurring traffic spikes at Disney's Hollywood Studios. Park planners correctly decided to make them an annual event, and anyone who has ever attended can verify that the difference in attendance is dramatic. As has always been the case, Star Wars is a huge money maker in virtually every form. The Star Wars Weekends concept quickly became the most popular time on the calendar for the third gate outside of the Osborne Family Spectacle of Lights. For years, rumors abounded that Disney was considering a larger Star Wars presence, and the home of Star Wars Weekends made the most sense.

Perhaps nothing would have come from this rumor if not for the long-standing paranoia Disney felt toward Universal Studios. Logically, no reason existed for this fear. Since the introduction of Universal Studios Florida, that park had barely caused a ripple in the theme park industry. In 2009, the third and fourth gates at Walt Disney World earned 9.7 million and 9.6 million in attendance. Universal Studios Florida only garnered 5.5 million.

The inability to bleed market share from Walt Disney World is precisely why Universal execs were willing to do something daring. They agreed to a contract with J.K. Rowling that Disney wouldn't. In the process, the Wizarding World of Harry Potter came to central

Florida, thereby fundamentally altering the history of many theme parks across the world. Disney wasn't about to build an entire area for an intellectual property they didn't own, no matter how child-friendly it was. Universal had every reason to do so. And that's how competition can change everything.

Fast forward to the end of 2015. Universal Studios Florida now has attendance of 9.6 million, whereas Disney's Hollywood Studios is at 10.8 million. One grew by more than four million tickets sold in six years. The other managed barely a quarter of that. Disney execs recognized their mistake. They'd rested on their laurels for too long, and the worst had happened. A theme park that wasn't one of theirs had become the most innovative and daring.

Anyone can connect the dots from there. Disney witnessed the appeal of a themed expansion land focused on a specific intellectual property. They saw it spike sales and draw newfound attendees to a previously barren park. They watched revenue increase dramatically. And the worst part was that it happened to a competitor instead of them.

Combine all of these facts with the recent acquisition of Lucasfilm. Disney now possessed a franchise that (arguably) exceeded Harry Potter in global popularity. They owned a Walt Disney World gate that had maintained a strange existence as a half-day park for a quarter-century. And they watched with horror as a rival turned a concept they'd rejected into a complete, viable business model.

In 2015, Disney execs offered a pair of seemingly conflicting announcements. On the one hand, the lucrative, wildly popular Star Wars Weekends would come to a close. Oddly, they chose not to announce this until after the 2015 event had transpired, meaning that Star Wars fans had no warning that it was their last chance to attend. Any damage caused by this bit of fan-boy sacrilege quickly became negligible.

In August of 2015, Disney CEO Bob Iger stated that Star Wars Land would become a reality at both Disneyland and Walt Disney World. The latter construction would occur at Disney's Hollywood Studios, thereby elevating the park to full-day status only 28 years after the fact. The timing of this news conflicted with the upcoming debut of the Wizarding World of Harry Potter at Universal Studios Hollywood.

The second Harry Potter expansion opened its gates in April of 2016. The expectation is that Star Wars Land will debut in 2019. When that occurs, two of the most beloved franchises in movie history will battle for the attention of theme park tourists in California and Florida. Independent of what happens next, these additions will elevate Disney's Hollywood Studios. Star Wars Land and the also announced Toy Story Land will tether the park to its intended concept of a movie-based theme park, an aspect that it has sorely lacked over the past 25 years.

One of the engrossing aspects of Hollywood history is that studio bosses are always stirring up controversy. They go to war with one another in awkward power plays that frequently cause more harm than good to both parties. In the case of Disney's Hollywood Studios, it's fallen victim to such conflicts since even before its inception.

First, Disney's paranoia over a new Universal Studios theme park in Florida led to the genesis of the third gate. Then, a questionable business decision by Ted Turner caused so many rifts at MGM that former and current employees of the company suffered from the ramifications for years afterward. Then, some of the new MGM corporate execs resented and lamented the existing contract for Disney-MGM Studios, which had yet to debut. They chose to renege on the deal, letting the court system decide its fate.

All of these issues fostered a negative stigma about the park. Even Disney cast members questioned how many resources they should invest in such a troubled project. When they did choose signature landmarks for their third Walt Disney World gate, those ideas were flawed and generally disliked. Even the one great idea that once served as the backbone of the park, the Great Movie Ride, suffered due to the corporate war between Disney and MGM.

Still, Disney's Hollywood Studios would continue to exist in a strange sort of half-life if not for an attraction called Star Tours. It built a relationship between Disney and George Lucas that eventually expanded into a popular getaway known as Star Wars Weekends. From there, Lucas developed enough trust in Disney as a business that he sold his life's work to them. This single business investment of $4 billion might prove itself as the saving grace of Disney's Hollywood Studios, which will finally evolve into a full-day park in coming years.

CHAPTER SIX

The Osborne Family Spectacle of Dancing Lights

Everyone knows that certain someone down the street who takes everything too far. They're just bad neighbors, and it feels like everything they do is specifically planned to drive you nuts. Imagine what it would feel like if those people who are the bane of your existence become national heroes and, later, icons of Walt Disney World.

The entire turn of events would be maddening, yet that's exactly what occurred for the people who lived on Robinwood Street. They're either the antagonists or the unintentional motivators in this piece. These unlucky residents of Pulaski County in Arkansas happened to live on the same street as a strong-willed millionaire. And they poked the bear when they tried to stop this businessman from doing something that pleased his only child, a girl called Breezy.

The details of the story are further proof that truth is stranger than fiction, even the meticulously crafted fiction of Walt Disney World. The instigator of all the less-than-neighborly issues was a man who earned millions performing human drug trials for salacious pills such as Viagra. Somehow, he became the greatest philanthropist in Arkansas, no small feat for a state where the Walton family of Walmart fame also resides. The causality for this unlikely evolution from instigator to benefactor started with something absurdly simple: Christmas lights.

Virtually no detail listed above makes sense, and that's why the story is so amazing. In 1986, an absentee father learned that his daughter wanted to spend time with him. They discovered a mutual love of holiday decorations. In the process, they disrupted the lives of countless residents of the capital city of Arkansas. Eventually, their shared passion progressed into such a divisive issue that the United States Supreme Court became involved.

After an unfavorable ruling appeared to end the lights show forever, the unlikeliest turn of events caused it to become one of the most famous Christmas displays in the world. Although Hollywood Studios has turned off the Osborne Family Spectacle of Dancing Lights for the last time, it's never too late to learn about the events that brought the wildly popular show to Orlando, Florida.

The year 1986 represents a different time for The Walt Disney Company. Only two gates at Walt Disney World and one at Disneyland were open at the time. Tokyo Disney Resort stood as the only international expansion, as Disneyland Paris remained six years away from its debut. Finally, although none of the cast members could have known it at the time, one of their central holiday selling points from 1995–2015 was beginning in the most modest of settings.

Jennings Osborne was the name of the central figure in this story. The Arkansas native came from humble beginnings, but he was the proverbial post-World War II American who pulled himself up by his bootstraps to become something greater.

By earning a degree in microbiology, Osborne proved that he owned one of the sharpest minds in his state.

Until the day he died, he remained a diehard supporter of his alma mater, the University of Arkansas at Little Rock, and he frequently hosted massive tailgate parties for their games as well as those of the flagship university in Fayetteville. From a young age, he felt unusually strong civic pride and constantly nurtured his Arkansas roots.

Putting his degree to excellent use, Osborne founded a company in 1968 that would turn him into a multi-millionaire. Fittingly named Arkansas Research and Medical Testing LLC, this business would grow into one of the premiere pharmaceutical research facilities in America. Their own site proudly proclaims: "Since 1968 we have conducted over 2,200 studies involving 28,000 volunteers and over 100 pharma/biotech sponsors."

Osborne always enjoyed bragging about the diverse drugs he'd had a hand in bringing to the market. Motrin, a pain reliever, and Viagra, a *ahem* stress reliever, both earned FDA approval thanks to human testing performed at the Arkansas facility. The entrepreneur's fingerprints were all over the expansion of his business, which is how he eventually garnered a net worth of approximately $50 million. The combination of a brilliant intellect, an explosive growth industry in big pharma, and some gutsy business choices turned the Arkansas native into one of the most respected microbiologists in the south.

A cliché exists with such successful business people, though. Family life tends to suffer when a person spends too much time at work. A founder of an eight-figure company is a workaholic by default. In the particular case of Jennings Osborne, he was lucky enough to marry a woman named Mitzi Udouj. She tolerated many long nights at home, and she suffered immeasurably in an attempt to bring a child into their life. Life punished her with a horrific five miscarriages before blessing her with a daughter.

Anyone even passingly familiar with the Osborne Family Spectacle of Dancing Lights knows about Breezy. Born Allison Brianne Jennings, she was understandably the joy of her parents' lives after so much pregnancy misfortunes. For obvious reasons, Allison Brianne was an only child viewed as a minor miracle by her doting parents. Even by only child standards, she was ridiculously spoiled, the proverbial apple of their eye, but work still kept her father away from home too often.

You can imagine the sense of devastation Jennings Osborne must have experienced in 1986 when his six-year-old girl requested an unexpected favor. The darling child he and his wife had struggled so much to bring into being wanted a simple thing from her daddy. She asked him to decorate their home with Christmas lights.

What she really wanted was to spend more time with him. Any parent would feel crushed by such an honest, earnest request. Even before the internet era, people wrestled to find the appropriate balance of home and work life. Breezy unintentionally informed Osborne that he had room for improvement in this regard. From the mouths of babes…

The lights display that first year stood out from the average displays of the era. There were approximately a thousand lights. Try not to guffaw at the thought of a legendary holiday exhibition such as the Osborne Family Spectacle of Dancing Lights featuring only a thousand lights. There are probably ten times as many Hidden Mickeys at the Hollywood Studios event.

From such humble beginnings, a shared love emerged. Mother, father, and daughter discovered their mutual enjoyment of shopping for the perfect Christmas lights and then putting them on the walls of their home. In year two, the holiday decorations escalated into something worthy of a drive for the other citizens of Little Rock.

The evolution of the Osborne Family Spectacle of Dancing Lights is well known by this point. In seven years, the once precocious six year old was now thirteen. Her adoring father had expanded their holiday decorations so much that the millionaire had to do something truly eccentric. He purchased the two adjacent homes in order to have more room for his holiday display. It's not quite as extravagant as Michael Jackson's Neverland Ranch, but owning two homes simply to have better December decorations is, well, wasteful at best.

What's involved in taking a thousand-light set of Christmas decorations into a three-house extravaganza? For starters, the Osborne clan added approximately three million lights. They added an entire nativity scene to emphasize the religious symbolism of the holiday. Then, they introduced a globe to highlight the placement of Little Rock, Arkansas, and Bethlehem, Palestine, the two most important locations for the event.

The circular driveway included rotating carousels of lights, which added visual wonder and pageantry to the proceedings. Since no holiday decoration is complete without a Christmas tree, the family constructed a seven-story tall tree right on top of their kitchen. The tree featured three different colors to enhance the holiday spirit, and the driveway alone included 70,000 lights, a factor of seventy more than the 1986 decorations.

If your wonder about their electric bill, you're not the only one. The family patriarch stubbornly refused to provide hard numbers about his joint venture with his daughter. At one point, the local utility company, Arkansas Power and Light, offered their insights on the topic. Without adding specific details, they acknowledged that the December utility bill for the lights show rivaled the average annual bill for citizens of Little Rock.

Two points emphasize the absurdity of the situation. The first is that the display took on a life of its own to the point that the Osborne family had to hire a full-time engineer. His sole job was literally to keep the lights on at the Osborne house(s). The other is that the display drained so many resources that one year when Jennings flipped the power switch for the first time, he blew the breaker for the entire neighborhood. And that was the moment of truth.

How would you feel if you were a neighbor of Jennings Osborne? It'd be like living next to Clark Griswold in *Christmas Vacation*, only ramped up several orders of magnitude. There were *three million* lights. Take the time to count how many lights you have stored in your holiday boxes. Then, perform some multiplication to understand just how massive an undertaking the Osborne family offering eventually became.

Today, many cities seem to have that one home whose holiday decorations are so festive that people drive across town to enjoy the view. In the early 1990s, the Osborne situation was still somewhat of a novelty. Given the larger-than-life persona of the Santa Claus-shaped microbiologist and the ostentatious nature of his light show, locals swarmed the accompanying roads to see the sights.

For the other people who lived on Robinwood Street and Cantrell Road, the entire situation was a nightmare. December is always a busy time for homeowners, whether they celebrate the holidays or not. Retail shopping and end-of-year get-togethers force people to travel more often than normal. Suffice to say that the logistics of the Osborne Family Spectacle v1.0 caused many trials and tribulations for the rest of the people living on the block.

Over time, people alerted the media that they needed as long as two hours simply to drive to the grocery store during December, the month when the lights were on display. It was a five-minute trip otherwise.

Traffic escalated by a factor of dozens when the rest of Little Rock headed to Robinwood Street to see the lights. And the lights themselves were no picnic. It was the holiday decoration equivalent of staring directly at the sun. Neighbors complained that they had to purchase thicker drapes and shades to reduce the glare caused by millions of Christmas lights.

Jennings Osborne was an accomplished businessman with a generous disposition and a terrific personality. He was also a terrible neighbor. When the other residents asked that he dial down the lights, Osborne responded by adding more. Three million more. It's simultaneously an obnoxious and hysterical response, although the people living there understandably found little amusement. The butting of heads put the microbiologist and six of his most opinionated neighbors on a legal collision course.

While a Rashomon aspect exists with regard to the he said/she said nature of the Osborne Christmas Lights lawsuit, all parties acknowledge a few pertinent details. For 35 days each year, Robinwood Street and Cantrell Road featured one of the most dazzling light shows in the United States. The people residing on Cantrell Road in particular suffered, because the three Osborne homes directly faced that road.

In addition to these complaints, a very serious problem existed. Due to the ridiculous level of traffic near the decoration event, cars couldn't get through easily. The other people living there rightfully worried about emergency situations. No early responders could navigate their way to a crisis situation in acceptable time, and every second counts in such situations. A delay of a few minutes could literally stand as the difference between life and death for people living on these streets. While everyone loves Christmas decorations, nobody wants to die because of them.

For his part, Jennings Osborne was resolute on the point that he owned three different houses on the block. He paid for everything out of his own pocket, including special dispensations to the utility board to ensure that his lights no longer endangered everyone else. They operated on their own transformer system. He felt that he'd been more than fair with his neighbors and remained resolute that he would not shut down something that pleased his daughter so much.

The war was on.

Half a dozen residents in the neighborhood filed suit against Jennings Osborne. They filed suit in Pulaski County, asking that the court file an injunction against their troublesome neighbor. They noted that Osborne had created six acres of headaches with no discernible benefits. His counterclaim was simple. "I do this to make people happy." The unbiased perspective of the matter is that both parties were correct. Osborne provided an inimitable service that enriched the lives of many Arkansans. In doing so, he wreaked havoc with the lives of the people residing nearby.

The suit itself was a slam dunk according to most legal analysts. It's simply an electric interpretation of the premise that your rights end where your friend's nose begins. As predicted, the court ruled against the multi-millionaire, albeit in a compromise. He could keep the lights on as long he limited the event to 15 days and kept to a strict schedule. Osborne could operate the festivities from 7pm to 10:30pm.

Most people would evaluate the situation as an amicable solution. The microbiologist raged over the ruling. After his attorneys examined the court ruling, they pointed out a loophole of sorts. He could just ignore the county courthouse, but it would cost him. To keep the Osborne Family Lights Spectacle thrilling his fellow citizens, he would have to pay a $10,000 fine. Given that the man bought two adjoining properties to boost the show's potential, this was a no-brainer. After only three days, his neighbors pointed out to the court that the entrepreneur had violated their terms. The actual legal documents are still online if you're interested.

As the Osborne clan kept the lights on, they appealed to a higher power, the Arkansas State Supreme Court. Their first attempt stood on more solid legal ground than a later one that fittingly used a religious argument to request removal of the injunction. Initially, the family attorneys argued that attempts by the neighbors to shut down the display infringed on Osbornes' First Amendment rights. The highest court in the state refused to negate the lower court ruling.

At this point, Osborne's lawyers attempted to entice the United States Supreme Court into accepting the case. They changed their argument. Many of the items on display celebrated the family's Christian faith. One of the most famous of them was a manger that created additional legal issues of its own a couple of years later.

Due to the deployment of such artifacts, these attorneys maintained that the lower court ruling to shut them down diminished Osborne's religious rights. He was unable to celebrate Christmas in his preferred manner. It was a desperate gamble that they only tried because everything else had failed. They determined that their best chance for a legal save was Supreme Court Justice Clarence Thomas, one of the most conservative members of the nation's highest court.

Thomas infrequently accepted such legal rulings, and Osborne had several connections with the man thanks to his friendships with soon-to-be Arkansas governor Mike Huckabee and current president of the United States Bill Clinton. Arkansas was an insular state, and Jennings Osborne was a beloved industrialist. Unfortunately, Thomas recognized what everyone else did. As much as people loved the family light show, it was infringing on the rights of their Robinwood Street neighbors.

The frustrated Osborne family patriarch later recounted, "I got 60,000 letters last year from people thanking me." He felt like the needs of the many should outweigh those of the few unfortunates living nearby. After all, he did many other things to bring positives into their lives. He hosted free barbecues in the backyard and never turned down a request to help anyone in need, even when they were suing him to shut down something he loved.

Bitter over the legal results, Jennings Osborne resolved to maintain Christmas decorations at the family home(s). He simply reduced them in scale to abide by the court decision. Otherwise, he would have eventually faced jail time. As it was, a judge gave him a 10-day suspended sentence. That left the family holding a ridiculous number of lights and other holiday paraphernalia.

They donated some of it to the city of Little Rock. Amusingly, this backfired a bit when civil rights advocates noted that the aforementioned manger shouldn't stand on display at a public facility. Such a move failed to separate church from state. Given everything else that had transpired, Osborne took this as the final insult for his generosity, noting that he could easily sell the piece for $30,000 or more.

As Jennings Osborne experienced the humiliation of public censuring for performing a kind deed for the community, his Christmas decorations went viral. News organizations across the country picked up the story. How could people sue a person for having one of the greatest holiday exhibitions in the United States? Isn't that the true definition of an American? In an era before the concept of going viral, the Osborne family lights show became a national curiosity virtually overnight.

One of the people who heard the Little Rock Christmas lights story was Bruce Laval, a vice president in what was then known as Disney's Theme Parks division. Laval assigned John Phelan, a show director at Walt Disney World, to make first contact. The Disney company had tasked Phelan with bringing a new Christmas pageant to Hollywood Studios or, as it was then known, Disney-MGM Studios.

The decision was understandable given that the other two Orlando gates, Magic Kingdom and Epcot, already offered majestic holiday festivities. Already perceived as the least of the three parks, Hollywood Studios needed to offer stronger entertainment to overcome its reputation as a half-day park. In watching the news, Phelan received inspiration.

The situation represented kismet at its finest. In Orlando, the leading family vacation destination in the world needed a new Christmas exhibition. In Little Rock, a frustrated entrepreneur owned literally millions of lights and decorations that the state court banned him from displaying. Everything seemed so perfect and yet an odd hiccup almost prevented the parties from forming a partnership.

Jennings Osborne was of course familiar with Walt Disney World. Everyone in America is, and most southerners vacation there at least once a decade. The Osborne clan had more money than the average southerner, which means that they were able to visit often. The close-knit family adored the place. The news that The Walt Disney Company wanted to host their light show *should* have put them over the moon.

Disney uses tongue-in-cheek naming conventions for their parks. When you enter Disneyland and Magic Kingdom, you walk down Main Street, a road that is relatable to everyone. Sustaining that philosophy, Hollywood Studios includes a place called Residential

Street. When Phelan pitched Jennings on transporting the Osborne show to central Florida, he announced his intention to locate it there.

Osborne, like a lot of casual Disney fans, didn't appreciate the distinction. What he heard was that his majestic family heritage of lights wasn't special to Disney. Instead, they would host it on a standard residential street. It was a Disney homonym that underscores the importance of capital letters, but those are difficult to emphasize in speech.

For his part, Phelan felt understandable confusion about the cool reception to his idea. He expected the Osborne family to embrace this turn of events as serendipitous. Had he not pressed Osborne further, the entire display could have easily never moved to Hollywood Studios. Thankfully, the parties worked through the miscommunication, saving the Osborne family lights for years to come. Phelan discovered his good fortune in a hilarious way. When he returned from his own family vacation, boxes of Christmas decorations were waiting in his office. Osborne had excitedly shipped them the moment they realized the show was moving to Disney World.

From The Walt Disney Company's perspective, the Jennings lights were a gift from above. The corporation notoriously planned and plotted everything. With the Hollywood Studios lights show, however, someone else had already done the initial shopping for them. The Osborne family shipped everything they thought Disney could use. The transported decorations required four 18-wheel Mayflower moving vans. And the price was right for any corporation: free.

The already rich Osborne clan had no need for money. They simply wanted to vacation at Walt Disney World over the holidays to visit the lights that they loved so dearly. The company agreed to host them free of charge onsite whenever they wanted during the holiday season. Disney feasibly could have spent millions of dollars on a light show. Instead, one fell right in their lap, and from Disney fans to boot.

With virtually everything transferred, the key decisions involved where to put the pieces. Some items such as the manger that proved controversial to the city of Little Rock wound up at the Italy Pavilion at Epcot. Other legendary pieces such as 100 angels, countless reindeer and Santas, and the epic 70-foot Christmas tree wound up at Hollywood Studios as signature parts of what was now called the Osborne Family Spectacle of Dancing Lights. If you've visited Walt Disney World during the holidays at any point over the past 20 years, you've undoubtedly enjoyed many of these decorations.

Because Disney is in the habit of plussing attractions, they never rested on Osborne Jennings' laurels. Instead, they constantly added new pieces to the collection. They did so partially to entertain the multi-millionaire. Cast members played a game of cat and mouse with the entrepreneur they grew to love, changing the location of some of his favorite items while introducing others. They knew that he would walk past every section, meticulously inspecting that it met his lofty standards, which were every bit as high as Disney's.

An adorable game even developed by happenstance. While sifting through the literal millions of items from the Osborne Lights boxes, they found a black cat. The Disney employees puzzled over the appropriate location of this oddity. Eventually, they quit trying and asked the curator himself. When queried, Osborne busted up with laughter, confessing that it was a Halloween decoration that his family accidentally misplaced in the wrong storage container.

The cast members ran with the joke, choosing to place the cat within the decorations. Osborne again experienced delight when he discovered the hidden treasure. From that point forward, hiding the Halloween cat became an annual tradition. The delightful game reinforced the perfection of the lights permanently residing at Walt Disney World.

When placed in their proper locations, the lights and other decorations cover 10 miles of Hollywood Studios. The wizards at Disney also found a way to disguise most of the extension cords needed to provide power to the breath-taking display. Those cords are approximately 30 miles long. Remember that when you look at your mess of tangled cords next time.

Putting up the lights was an ordeal. Disney spent 20,000 man-hours each year on the project, the equivalent of 500 40-hour work weeks. The power required to keep the lights on was staggering. Disney estimated they used 800,000 watts during the six weeks that the show ran each year, from mid-November until early January. Disney mitigated the costs of the electricity by persuading Sylvania to sponsor the display in 2005.

The year prior to the sponsorship, a major change occurred. After 10 years of operation on Residential Street, the demolition of that area to make room for Lights! Motors! Action! Extreme Stunt Show necessitated relocation of the Osborne Family Spectacle of Dancing Lights to the Streets of America, a title that likely also confused Jennings Osborne at first.

Originally, fans worried that the show might go away. Those fears proved unfounded at the time. As late as 2011, just before tragedy befell him, Jennings noted in an interview that people caused too much fuss about the future of his exhibition. He noted that Disney had just extended the contract yet again, and he expected them to continue to do so.

But the innate largesse of the microbiologist eventually proved his undoing. It also had help. No man who loves barbecued ribs that much is assured of a long and healthy life. On July 27, 2011, Jennings Osborne died of a heart ailment. He was 67 years old. By that time, he'd spent almost half of his life enriching the lives of others with his lights show.

Even before Osborne died, however, his natural generosity had cost the man his fortune. He paid for holiday lights for people in more than 25 cities across the country. He frequently provided financial aid to people struck by tragedy. He even paid for the funeral arrangements when locals informed him of people who couldn't. Even the barbecues that increased his chances of heart problems earned enough in charitable donations to feed thousands of indigent people in his state. He was a hero to his very core and the very definition of someone who would literally give a stranger the shirt off his back.

After his death, the long-rumored financial woes about Jennings Osborne proved correct. The court system, unkind to him even after his death, ruled that his estate was several million dollars in debt. It was a stunning financial collapse, particularly in light of the fact that he sold his company for $24 million in 2004. Facing bankruptcy, the family was left with no other choice. They eventually had to sell the three houses from which the Osborne Family Spectacle of Dancing Lights originated.

In 2015, Mitzi and Breezy received additional bad news. The Walt Disney Company announced that the Star Wars Land and Toy Story Land expansions required part of the tract ordinarily reserved for the Osborne Family Spectacle of Dancing Lights. People had already deduced the implication of the construction work at Hollywood Studios even before the announcement.

On September 11, 2015, Disney confirmed that they would discontinue the beloved Osborne lights show on January 3, 2016. In December of 2016, for the first time in 30 years, the Jennings clan will not display a lights show at either their former home on Robinwood Street or their adopted home at Hollywood Studios. Truly, it's the end of an era.

Perhaps the saddest part of the closing is that the family who only recently lost their patriarch is now forced to watch his legacy die. And that comes on the heels of his overly generous nature leaving him millions of dollars in debt, meaning that the homes that embodied a key part of his legacy are no longer owned by the family.

Jennings Osborne was an innovator, a philanthropist, and an American original. His impact on the citizens of Little Rock, Arkansas, once appeared to be his legacy. Then, his breathtaking exhibition of dancing lights transferred to Walt Disney World. At that point, he became a signature part of the annual holiday tradition of many families. Millions will miss the Osborne Family Spectacle of Dancing Lights, and Disney would be wise to take those millions of lights out of storage and put them back on display sooner rather than later.

CHAPTER SEVEN
Animal Kingdom

One of the visions Walt Disney possessed for his upcoming theme park is largely ignored by history. During his quest to reinvent the amusement park as something bigger and better, he sacrificed some of the initial plans. Part of the explanation was financial. As the situation played out, Disney already had to barter many of his prized possessions while nearly bankrupting his company. Some of the other reasons were pragmatic. Certain ideas are better in theory than execution.

In the end, the primary reason why some of Walt's grand ideas fell by the wayside was a lack of resources. His construction team built Disneyland in almost exactly one year. Any idea they couldn't block out and bring to fruition quickly was dismissed. That's why one of his most daring notions fell by the wayside.

Walt Disney wanted more than just an amusement park. He also wanted a zoo, a place where children could interact with animals in a safe environment. Infrastructure challenges sabotaged this dream, which led to compromise in the Jungle Cruise, an *African Queen*-esque river ride featuring permanently immobile animals.

While Jungle Cruise is one of the greatest theme park creations of all time, it's also a perversion of Walt's vision. He wanted a zoo. Even after he passed away, his loyal team of Imagineers and their later successors remained steadfast about staying true to Walt Disney's core values. They carefully recorded and categorized all the ideas he postulated over time.

Less than four years after Walt's death, the company founded the Walt Disney Archives. Their goal was to protect his legacy and, whenever possible, bring his futuristic plans into reality. Walt Disney was a man ahead of his time, and that's exactly why some of his dreams were a functional impossibility in his lifetime. It's remarkable that he achieved as many as he did. Even with all his staggering accomplishments, many of Walt's fertile ideas germinated long after he died.

The most noteworthy of these was an interactive environment for kids to learn about nature. It was intended as a place where kids could befriend animals while receiving an education about their surroundings. Walt Disney desired it in 1955, but his staff couldn't find a way to protect either the children or the animals. A natural habitat for a single species is trying enough. Building dozens of them for various species wasn't feasible in the 1950s.

Four decades later, a team of Imagineers took on the challenge. They embraced their founder's bold vision. They started work on what would become the fourth gate at Walt Disney World, a place where children could play with fuzzy animals and less fuzzy ones, too. They built a series of structures that brought stability to hundreds of animals, honoring their natural environments. And these Imagineers did so in a way that children could watch, play with, and learn from the beasts.

While another gate claims the title of Magic Kingdom, the true wizardry in park design takes place at Disney's "newest" theme park. Here is a detailed look at the history of the planning of Animal Kingdom, a glimpse into what-might-have-been if one of the ideas on the drawing board had been built, and a brief look at the park's future.

After the successful launch of Disney-MGM Studios in 1989, then-CEO Michael Eisner felt a sense of urgency in sustaining The Walt Disney Company's momentum. His grand ambition at the time was to add a fourth gate quickly. By 1990, Disney-MGM Studios was regularly in the habit of turning away potential guests within hours of park opening. Eisner believed that the constant influx of traffic at a perceived half-day park, the recently launched third gate, proved the necessity of a fourth one.

While Eisner's performance as CEO was already controversial, even his detractors acknowledged that the corporate leader respected Walt's ideas. As his staff tossed around ideas and debated their feasibility, Eisner returned to one of Walt Disney's greatest incomplete ideas. He wanted to build a theme park that was functionally a zoo as well. Those same feasibility debaters were lukewarm to the idea. Had the premise proved viable, Disney himself would have attempted it 40 years before. He had the original Imagineers spit-balling ideas for him. If they couldn't do it, the pervasive belief was that it wasn't possible.

Undeterred, Eisner picked a lead Imagineer to investigate the potential of a Disney theme park zoo. That man's name was Joe Rohde. Some people, myself included, consider him the greatest working Imagineer today. In 1989, however, his legend was just in the earliest stages. By his own account, Rohde transitioned from a high school director to a low-level Disney employee. After several years with the company, he hadn't left an imprint of note. Then, he garnered a plum assignment working on the Adventurers Club portion of Pleasure Island. Through this project, he earned the trust of his superiors.

When the time came to proceed on the fourth gate, Eisner favored Rohde for the zoo study. Rohde would later note that his principal advantage in his early days with the company was his willingness to speak his mind during staff meetings. Had he been quieter by nature, the man with the legendary earring would remain an anonymous low-level employee.

During his investigation, Rohde met with a few zoologists. A myth about this process that isn't far from the truth is that he kept meeting with people until he found someone who believed it was possible. He first reached out to the executive director of the Bronx Zoo at the time, Bill Conway. This conservationist was one of the true heavyweights of the profession. He would eventually anchor the Wildlife Conservation Society. It wasn't that he was indifferent to the idea. It was more that he already had a full-time job, so he didn't have the free time to plan for a theme park that might never come to fruition.

Conway placed Rohde in contact with a man who would become seminal to the development of Animal Kingdom. His name was Rick Barongi, and he spent 40 years working in the animal conservation industry before retiring in 2015. If Rohde is the unofficial father of Animal Kingdom, Barongi is a favorite uncle, at a minimum. Rohde warmed to him due to his positive attitude. In situations where other zoologists told the Imagineer that something was impossible or, at least, improbable, Barongi would take a different approach. His favorite reply was, "I don't see why not." The entire Disney Imagineering institution is built on that sort of optimism.

The two men worked together to identify potential sticking points in a theme park zoo. Disney's least popular theme parks received millions of visitors each year. Many animals are so shy that they spook at the sight of a single human. How would Disney build an entire habitable environment for hundreds of different species, some of whom wouldn't react well to the prying eyes of mankind?

Rohde and Barongi embraced this challenge. They investigated thousands of species, deciding which ones would play well not just with humans but with other animals. A single misstep would have an unintentional but horrific result. Children enjoying a day at Disney's zoo might accidentally witness a cute, Bambi-ish animal devoured by a less cute, much hungrier foe. The circle of life sounds lovely when Elton John sings it. In execution, it's gruesome and less than family-friendly.

The Imagineer and conservationist invested much time solving these issues. As Barongi later recounted, his quest became a source of amusement/bemusement amongst his peers. They fell squarely into two camps. Some of them felt that the very idea of a Disney theme park zoo was an obvious impossibility. Every second of his life he squandered on the foolish endeavor was a mistake. The rest felt envious about his situation. They had complete confidence that Disney's resources would allow them to build the planet's most mainstream zoo. They wished they could join Barongi in the venture.

The latter philosophy, the belief in Disney, proved to be the correct one. And that envy about participating in the process later wreaked havoc with zoos across North America as the opening of Animal Kingdom approached. You'll understand why in a moment.

Rohde approached Barongi for the first time in 1990. The park opened in 1998. You can fill in the dots from there about how smooth the construction of Animal Kingdom was. To be fair, not everything was Disney's fault. They always understood that the planning phase for such a monumental endeavor could take several years.

To build the best possible landscape as well as a series of habitable environments for diverse animals, the company would need to perform an unprecedented amount of due diligence for a theme park. Over the course of just a couple of years, they interviewed hundreds of zookeepers. They obviously wouldn't invest so many man-hours and resources into the project unless they felt confident it would proceed.

What no one could anticipate was just how laborious the task would be. Multiple years passed with little tangible progress to show. At the start of 1991, Disney still attempted to keep the project secret from most of its own employees, even as they interviewed zoologists and other animal behavioral experts. Word eventually spread, of course, as non-Disney employees queried on the matter would network with friends gainfully employed there.

Rohde himself expressed confidence that Animal Kingdom would become a reality soon. He recommended Barongi to the higher-ups at Disney, and the conservationist finally became a cast member in 1993 after several years as a secret consultant. His title included oversight of the company's animal operations. This gave him a job beyond the still in-doubt Animal Kingdom, as he also handled tasks at Discovery Island and The Living Seas. In this manner, Disney protected him in the unlikely event that Animal Kingdom was delayed.

In 1994, Animal Kingdom was delayed.

More than 20 years ago, Disney's theme parks once again proved susceptible to two things utterly beyond their control: death and taxes. The taxes portion is a cheeky way of acknowledging that the American economy struggled mightily that year. Whenever this occurs, many citizens cut out luxury purchases such as vacations, which directly impacts Disney's bottom line.

On top of the sluggish economic performance that year, Disney lost one of its titans. Frank Wells was the president of The Walt Disney Company from 1984 until his death in 1994. Wells took an ill-fated helicopter ride, and died in the crash. Unfortunately for the theme park zoo project, Wells was also one of its strongest champions. In his absence, a power vacuum occurred within the company. Simultaneously, Animal Kingdom lost a key supporter for an expensive project. At a time when money was tight, the lack of an ardent proponent proved to be a crippling setback.

Disney wasn't done with the internal turmoil, either. Also in 1994, Michael Eisner sandbagged Jeffrey Katzenberg, the man most likely to replace Eisner as CEO. The cause of this was, once again, the death of Wells. The gist is that Disney wound up spending another $270 million to buy off Katzenberg. Before they did so, they wasted millions of dollars in legal fees arguing against his claim. In short, one helicopter crash damaged Disney in multiple, profound ways. Animal Kingdom as a concept suffered the consequences of all these issues, causing it to sit on the backburner for another year.

On June 21, 1995, six years after originally discussed and three years prior to its actual opening, Animal Kingdom was confirmed. Buoyed by a 20 percent increase in operating income over the previous six months, the company felt the time was right to upgrade Walt Disney World. Thirty years after Walt Disney shrewdly purchased 27,258 acres of Florida swampland on the down low, his successors at the company devoted a big chunk of them to this park, the proverbial fourth gate. The public relations department trumpeted the $750 million expansion for its revolutionary nature.

Disney emphasized the park's strongest selling point. Guests of all ages would have the ability to interact with many of the wildlife species they'd watched on their favorite television programs. Disney planned to import animals from parts of the world few Americans ever visit, giving people unprecedented access to these noble creatures.

Michael Eisner, never one to shy away from publicity, saw this press conference as his moment of glory. From his perspective, Walt Disney World now offered four parks. Half of them had his fingerprints all over them, making him the modern-day Walt Disney. While his detractors would blanch at the thought of this argument, there's also validity to it. Disney-MGM Studios and Animal Kingdom were both his initiatives, and 18 years after the latter park debuted, they remained the two most recent gates on site.

The belief is that more than 1,700 species of animals reside at Animal Kingdom, although the suspicion is that this is a conservative estimate. Once Disney authorized the capital outlay for a new park, the quest became simple. They had to find enough people to work as overseers for countless species, many of which had never lived in North America before. And the job was made that much more difficult by the mercurial climate of central Florida.

Disney proceeded in a way that only the most powerful corporations can. They threw money at the problem. They could do this for a simple reason. As Rohde has mentioned over the years, his small team of Animal Kingdom designers was slender in its early days. While the project was primarily theoretical, he could keep the budget lean. That meant communicating with lots of consultants, none of whom worked for Disney. Prior to its confirmation, Animal Kingdom cost little to the massive bottom line of The Walt Disney Company. Once announced, it was noteworthy for having too few employees.

Rohde and Barongi leveraged all the contacts they'd made during the exploratory phase. They poached a jaw-dropping number of the world's finest zookeepers. Disney briefly caused a panic in the zoo community. Long-standing facilities suddenly found themselves without their best and most experienced staff members.

Everyone who was anyone in the industry headed to Walt Disney World to take up permanent residence. Meanwhile, sixty-nine different North American zoos had to put up "Now Hiring" signs. Disney's gain was a bloodbath to the zoo community as a whole.

As for how they accomplished the smooth interactions between complicated animal ecosystems, that's a topic worthy of an entire chapter on its own. What's undeniable is that Animal Kingdom was a project Walt Disney's people once considered so impossible that they chose artificial animals instead. Forty years later, a team of many of the world's finest conservationists pulled off the task so deftly that the achievement seemed effortless. It wasn't, of course, but that's how fine a job they did.

One of the most legendary days at Disneyland was also arguably its worst: opening day. The lingering memory of that debacle drove Imagineers and park operators to consider everything prior to the opening of a new location. In the case of Animal Kingdom, Disney could only control so much. The media grew obsessed with the ambition of the project.

Disney announced the opening date of April 22, 1998, which was fitting. It was Earth Day. Reporters requested a historic number of press credentials. Few of them believed that the company could pull off such a difficult task. Everyone wanted to see how well the animals, many of which were natural enemies, could co-exist. The same was true of normal theme park tourists. They too were curious about the new endeavor that was supposedly half-amusement park and half-zoo. On paper, it appeared to be the most novel Earth Day ever. Plus, Disney was pulling out all the stops for its latest offering. Opening day visitors would be met by performing African bands, given a grand opening lithograph, and taken along a path comprised of rose-petal confetti.

The gates were due to open at 6am., an hour earlier than the standard operating hours Disney had previously announced. They did this in anticipation of massive crowds, but even long-standing cast members expressed surprise at what happened next. The park capacity at the time was somewhere between 15,000 and 22,000, depending on which media report was reliable.

Within 75 minutes of Disney's opening the gates to Animal Kingdom, the park had already exceeded maximum capacity…and by *a lot*. A Disney spokesperson confirmed to the news services that they counted paid attendance in excess of 28,000 during this brief window when the gates were open.

The scary thought is that this total doesn't reflect two other types of guests. Approximately 5,000 reporters used their credentials during this period. Disney employees had sagely tipped them to get in line as soon as possible. Otherwise, they would have missed the story entirely due to the quick closing of the gates.

The 33,000 guests mentioned above also didn't include one other group. Annual pass holders to Walt Disney World didn't count as part of the paid tickets the company reported. It's fair to say that Animal Kingdom exceeded capacity by as much as double on its opening day. Even using the most favorable numbers, at least 10,000 too many visitors took part in the festivities. And all of this happened by 7:15am. It was a bad day to sleep in.

What's memorable about the day beyond the shocking lack of crowd control was that the opening of Animal Kingdom occurred without incident. The park was full of broadcasters, with many of the media credentials going to television and radio services broadcasting live from the newest theme park at Walt Disney World. Cast members cleverly roped off sections of the park to give all these media outlets free reign during their broadcasts. They rarely interfered with standard guests.

The animals themselves were on their best behavior. To the credit of all the conservationists Disney had poached from popular zoos, the habitats they built for the fledgling animal communities held up under the stress of continuous interactions with other animals as well as humans. This was surprising to all involved. In the build-up to the grand opening, the media had savaged Disney over reports of animal fatalities. Whether those issues were overstated or the company's newest cast members learned from the incidents is up for debate. What's inarguable is that everyone got along on day one.

The only true problem was avoidable, for reasons you'll see in a moment. The members of the press seeking to accentuate potential controversies at Animal Kingdom felt annoyed by the lack of animal issues. Disappointed that their planned headline wasn't available, they found another source of discontent.

Some of the opening day guests simply didn't *get* Animal Kingdom as designed. They thought of it in simplistic terms as a theme park with animals. These visitors rushed to the rides in parts of the park such as DinoLand U.S.A. and Asia's Kali River Rapids. Once they'd ridden the few true rides at Animal Kingdom, they expressed confusion that the park had so little to do. Meanwhile, the guests who interacted with the animals felt a sense of awe at the achievement of Animal Kingdom. Those people wound up in articles written by optimists. The people who left the park by noon, annoyed by the crowds and lack of E-ticket rides, wound up starring in articles written by negative media outlets.

The truth was somewhere in the middle. On opening day, Animal Kingdom went as well as Disney could have hoped. It was clearly a source of intrigue for both the local Florida community and the rest of the country. The park was lacking in some key regards, though. This would become true in the coming years. After starting with solid park attendance of 8.6 million during its first full year in 1999, Animal Kingdom fell in popularity each of the next three years.

History shows that the first public confirmation of Animal Kingdom was a bit ambitious. What it lacked in terms of rides may have cost the park during its early days. Also, the continuing perception of this issue is an idea Disney park planners fought right up to the day Pandora: The World of Avatar opened.

Eisner himself stated that Animal Kingdom would be a "celebration of animals that ever or never existed." That's because Disney feared that their newest park could exist as its own entity without a bit of aid. What they worried about was that the basic zoo concept wouldn't seem special enough for demanding theme park tourists, customers who expect more from the Disney brand.

In order to elevate the concept and also provide a few novel attractions, Disney planned to add mythological and prehistoric animals to its menagerie. The linchpin idea was that the fictional rides as well as the ones from millions of years ago would provide different storytelling avenues for Imagineers. That way, they could protect their interests while hedging their bets if the animals didn't prove enough of a draw.

Lost in the annals of history is Beastly Kingdom, which became Camp Minnie-Mickey. As bizarre as that sentence reads, it's true. While Expedition Everest stands today as the notable entry in the fictional beings portion of Animal Kingdom, the first announcement for the park explicitly stated that there would be many of them. In execution, none of note existed during the first eight years that the park was open.

What was the source of this disconnect? Beastly Kingdom wasn't a part of Animal Kingdom for a lack of trying. To the contrary, Imagineers invented several memorable concepts for this part of the park. The grand ambitions for Beastly Kingdom would have involved "realms" for the good and evil creatures from folklore.

Attractions such as Fantasia Gardens and Quest of the Unicorn would highlight the side of the angels. The former ride, obviously based on the movie *Fantasia*, would have combined music and the animals appearing in the film. Think of Many Adventures of Winnie the Pooh, only with crocodiles and hippos from "Dances of the Hours." Those are obviously real animals, but fake ones would have comprised other parts of the ride. Chief among them were centaurs, fauns, and pegasi from "Pastoral Symphony." Quest of the Unicorn was to be a labyrinth. Clever children who reached the middle would get to meet with the legendary unicorn that lived there.

On the dark side of Beastly Kingdom, theme park tourists would interact with the more foreboding mythical creatures of legend. Nearly two decades before a Ukrainian Ironbelly lorded over the streets of Diagon Alley at Universal Studios Florida, Disney wanted to let there be dragons. Their planned evil realm would feature Dragon's Tower, a vault of gold akin to the one seen in *The Hobbit: The Desolation of Smaug*.

The only thing preventing theme park tourists from reaching the treasure would be an avaricious dragon. This fire-breather would function as the final boss standing in the way of Scrooge McDuck-level riches. It was as ambitious as it was prescient. The next time you ride Harry Potter and the Escape from Gringotts, you should remember that Animal Kingdom could have easily beaten the Wizarding World to the punch.

What kept Beastly Kingdom from becoming a crucial part of Animal Kingdom? As the project advanced, park planners determined that they couldn't meet the original launch date for the fictional realm. They quickly devised a Plan B. Beastly Kingdom would become the first expansion of Animal Kingdom, the Phase II if you will. Once they'd proven that the theme park zoo concept was viable, they could then Disney-fy it by adding these new mythical attractions. They'd lead with the zoo and then bring the Magic Kingdom aspect into the equation at a later date.

As so often happens with plans for an indefinite time in the future, they fell by the wayside. The expense of Beastly Kingdom would be tremendous, and it would occur soon after Disney had finally finished accruing construction expenses for Phase I. As much as Disney loved the idea of a village residing under the shadow of an intimidating Dragon Tower, executives had to make hard choices.

The original $750 million projected budget for Animal Kingdom had spiraled out of control, as history indicates usually happens with Disney theme parks. As opening day approached, Michael Eisner faced a seemingly impossible choice. He had blueprints for two parts of the park, both of which he'd personally promised when he announced the intention to create Animal Kingdom.

One of them was Beastly Kingdom. The other was DinoLand U.S.A. You already know how this plays out. The difficulty is in understanding why he selected one of the most often-criticized parts of any Disney theme park over one with seemingly limitless potential. The explanation is one that seems comical in hindsight, but it's also the very business model Disney has used to dominate the theme park industry.

In May of 2000, Disney would release their most ambitious animated project in decades. The project was simply named *Dinosaur*, and it would be the company's attempt to prove that they could best Pixar at their own game, computer animation. Controversial director Paul Verhoeven had first pitched the project in 1988, and it had gestated behind the scenes at Disney for several years before proceeding in the late 1990s.

Eisner recognized an opportunity for synergy, the tethering of a potential film franchise to a newly incorporated theme park land. This strategy had worked many more times than it had failed for Disney, so his decision was understandable at the time. Unfortunately for him, *Dinosaur* lacked the one thing that differentiates Pixar from its derivatives: a quality story. It disappointed at the box office, and the glorified carnival region of Animal Kingdom continues to lag behind the other, more engaging parts of the fourth gate. Simply stated, the park needed a hook, and *Dinosaur* never delivered that hook. Beastly Kingdom possibly would have.

At the time, even Eisner presumed that the selection of DinoLand U.S.A. over Beastly Kingdom was merely a short-lived delay. During his opening day speech at Animal Kingdom, he stated that the new park was "a kingdom ruled by lions, dinosaurs and dragons." While waiting for the introduction of Beastly Kingdom, one that would never occur, Disney execs added what they considered a temporary placeholder.

Named Camp Minnie-Mickey, it was a character meet-and-greet with a couple of shows. The less popular one was Pocahontas and Her Forest Friends, while the one that maintains popularity to this day is Festival of the Lion King, the only attraction from any of this that has stood the test of time, and which now resides in a specially built theater in the Africa section of Animal Kingdom. Meanwhile, the idea of Beastly Kingdom fell out of favor once Islands of Adventure at Universal Studios Florida aped the idea.

The legacy of Beastly Kingdom is understated. Artists started designing elements of the park long before Eisner killed the Phase II expansion. They unfortunately named one of the parking lots Unicorn. The exterior entry display also features a dragon's head. Throughout the park, other dragons, unicorns, and other mythical creatures continue to make unlikely appearances on sign posts, logos, and other corporate paraphernalia. These are cheeky references to what might have been had Eisner favored creatures of legend over a film with a mediocre film rating of 6.4 on IMDb.

The only constant is change. The maxim is tired, but the underlying philosophy is sound, particularly within Disney's corporate offices. Innovation equals revenue, and money drives their business as much as creativity. Recently, Animal Kingdom and Hollywood Studios have jockeyed for a strange position. Neither wants to stand as low gate on the Walt Disney World totem pole.

Their traffic numbers are virtually indistinguishable. From 2010 through 2014, Hollywood Studios claimed 49.64 million visits. Animal Kingdom's numbers during that timeframe were 50.07 million. In 2011, park planners attempted something unprecedented to differentiate Animal Kingdom. After watching their failed Harry Potter negotiations lead to a globally popular license reinvigorating Universal Studios Florida, Disney tried once more.

On September 20, 2011, the Disney blog's editorial content director, Thomas Smith, announced what seemed like the most historic news since the introduction of Animal Kingdom in 1998. The fourth gate would receive an update that had virtually nothing to do with animals. Instead, it would harken back to the fictional beasts once considered. Animal Kingdom would add Pandora: The World of Avatar. The James Cameron film had dazzled movie lovers with its breathtaking visuals, ultimately becoming the number one movie of all-time domestically and globally. It was *the* film story of 2009 despite it not being released until the last few days of the year.

Not quite two years later, Disney decided that they wouldn't make the same mistake twice. After missing on Harry Potter, they would bank on the wizardry of James Cameron, presuming that the combination of Disney's Imagineering power and his storytelling would dazzle theme park tourists.

The novelty of this relationship was that Disney hadn't released *Avatar*. That windfall went to 20th Century Fox. Disney committed an entire block of their Project X land to an intellectual property they didn't own. Sure, they'd built rides for film franchises such as *Star Wars* and *Indiana Jones*, neither of which they owned. But they'd never gone all-in with an entire region of park space, one of their greatest commodities. The explanation was simple. One of the stumbling blocks with Harry Potter was that Disney didn't want to commit to such a project. They quickly appreciated the error in judgment. *Avatar* would afford Disney some redemption.

While Walt Disney could build Disneyland in a year, the technological hurdles of implementing a 3D world as a theme park land proved difficult, even to the geniuses in Disney Imagineering. Construction didn't even begin for more than two years. Builders eventually broke ground on Pandora in January of 2014. The following year, Disney offered more concrete details at their annual D23 Expo. They confirmed two attractions, Flight of Passage and Na'vi River Journey, both of which would simulate key sequences from the first movie. And that's an important note about the delays.

Part of the problem is that James Cameron announced a pair of sequels to *Avatar* in 2010. Their scheduled release dates were 2014 and 2015. The years came and went without new movies, and Disney suffered delays as well. Finally, in 2017, Disney debuted a land that promised to be the most forward-thinking expansion in the history of theme parks.

Disney's bold move to refresh Animal Kingdom has started off strongly, with rave reviews of Pandora in its opening months. In particular, Flight of Passage has been lauded as one of the most innovative experiences in theme park history. Will the world of Pandora enable the fourth gate to dispel the notion that it's a half-day park? It may be too early to tell since it's still so fresh to park visitors, but early signs are positive.

CHAPTER EIGHT
Beastly Kingdom

The best land at a Disney theme park never got made.

Don't take my word for it. Many of the Imagineers involved with designing and constructing the blueprints for this project share my opinion. Some of them felt so betrayed by the Disney CEO's decision not to build the land that it became one of the earliest rallying cries against Michael Eisner, whose tenure with the company ended acrimoniously.

As *The New York Times* chronicled in 2005, Eisner's departure earned nothing more than a "one-page retrospective in the company newsletter." For a man with a 21-year tenure as CEO to walk away with such little fanfare spoke volumes about the situation. Eisner exited Disney with a whimper rather than a bang.

A civil war with an actual member of the Disney family caused Eisner to lose his hold on the company he'd led since 1984. One of the reasons Roy E. Disney stepped down from the company his uncle founded and his father ran was that he'd lost faith in Eisner. Under Eisner's watch, the Disney company had added only two new gates to Walt Disney World plus one at Disneyland. The Orlando theme parks were considered half-day parks while the new Disneyland expansion, Disney California Adventure, was widely regarded as a bust. And let's not forget that "cultural Chernobyl" thing with Disneyland Paris, either.

The park that infuriated diehard Disney fanatics, however, is one that's probably the best of four. Disney's Animal Kingdom is a hallmark achievement in themed park construction. Disney triumphantly blended the popularity and family appeal of a zoo with a collection of wonderful rides that continue to entertain to this day. Only Disney would have the daring to place thousands of animals within stampede distance of thousands of tourists yet somehow safeguard the lives of creatures and humans alike. They did this while also controlling one of the most challenging aspects of a zoo: the smell. Even the opening day reviews for Animal Kingdom expressed amazement that Disney had overcome every issue that stems from a visit to the zoo.

What few people outside the company realized, however, is that all of those small and large victories came at a cost. The strongest potential draw at Animal Kingdom wasn't a part of opening day at the park, at least not in a meaningful sense. Corporate execs promised theme park tourists and Imagineers alike that the construction would occur as part of the second phase of the park. That promise was also broken by Eisner. In the end, *the* best themed park concept in the history of the company went unbuilt, at least at a Disney theme park.

This is the story of how Disney park planners caught lightning in a bottle, inventing one of the most marketable themed lands of all time. It's also the story of how and why Disney's upper management spit the bit, hurtling down a linear path of terrible decisions that eventually cost the world a chance at something special. This is the story of Beastly Kingdom, which is the most popular Disney destination in some alternate dimension. In our realm, it's sadly a tale of what might have been.

"Welcome to a kingdom of animals...real, ancient and imagined: a kingdom ruled by lions, dinosaurs and dragons; a kingdom of balance, harmony and survival; a kingdom we enter to share in the wonder, gaze at the beauty, thrill at the drama, and learn."

The above is a snippet from Michael Eisner's grand opening speech at Disney's Animal Kingdom. You should pay attention to that one creature type listed in the same breath with lions and dinosaurs. You've seen those lions during a thrilling expedition on Kilimanjaro Safaris, and you're intimately familiar with DinoLand U.S.A., which has a well-earned reputation as the junkiest themed land at Walt Disney World. But where are the dragons, you wonder?

Was Eisner referencing the Komodo dragons that inhabit Asia's Maharajah Jungle Trek or the bearded dragons in Conservation Station? No, of course not. The CEO of the corporation used his introductory remarks to set the stage for a later expansion, one that hadn't quite made the cut for the first phase of Animal Kingdom. The exclusion of this realm was the harshest budget-based reduction of the ambitious plans for the fourth and (thus far) final gate at Walt Disney World.

In building the world's most impressive functional zoo and natural habitat, Disney battled with their ledger sheet from the inception of the project. When guests think of Animal Kingdom, the first thing that springs to mind is the animals. What few folks contemplate is how much Disney has to pay for their food, health, and daily maintenance, not to mention keeping natural predators away from potential prey. It's a costly endeavor.

Your natural inclination is likely to dismiss this expense, recognizing that the price of admission at Animal Kingdom in combination with food and merchandising revenue pays the bill for Disney's animal upkeep. That's absolutely correct today. It wasn't the case in the mid-1990s, though. Back then, the corporation was trying to launch Animal Kingdom. In order to achieve this goal, they had to entice many of the finest zoologists in the world to leave their current jobs behind and join Disney. Simultaneously, they asked these experts for suggestions about the beasts the park would need to import to boost its profile and reputation within the theme park industry.

None of the above is cheap. Disney had to pay for it all, and they wouldn't have an evergreen revenue stream for several years. This opportunity cost of park creation explains why every time Disney builds a new theme park, the output isn't as great as early buzz had indicated. Also, the corporation inevitably winds up cutting costs at other parks to reduce overhead. In the case of the original Disneyland, Walt Disney famously sold his "dream house" to finance the project.

Once theme parks are operating and drawing in consistent crowds, they're solid breadwinners. Getting them off the ground requires the clearing of a ludicrous series of financial hurdles. In the case of Animal Kingdom, the one hurdle they could never clear was their best idea. It was a land that would have proudly hoisted a sign stating, "Here there be dragons!"

Zoos have always maintained a steady popularity in terms of tourist visitation. In building Animal Kingdom, Disney faced a seemingly insurmountable problem. They had to build a bigger, better zoo, and it held additional challenges, too. The new gate at Walt Disney World had to sustain multiple habitats of creatures, many of whom were natural enemies. Many hundreds of inhabitants would call Animal Kingdom their home.

Park planners rightfully wondered what would happen if they struggled in their early days as zookeepers. What would happen to their latest park if the early buzz were negative? After the perceived misses at Disneyland Paris and Disney/MGM Studios, they couldn't afford another high-profile failure, even if they didn't view either of the other two parks in that light. The media did, and their constant negative stories about each business became a self-fulfilling prophecy of sorts.

To avoid a third mistake, Disney execs understood that they needed an ace in the hole. That ace was a premise that tested well with theme park tourists. Disney has been the master of the survey for decades now, and they meticulously sampled the population during the early phase of Animal Kingdom planning. When they pitched the idea of a Disney zoo of sorts, complete with creatures who weren't indigenous to North America, guests were largely positive.

The idea of a land of dinosaurs did well in the wake of the Jurassic Park franchise. In fact, in the year prior to the opening of Animal Kingdom, *The Lost World: Jurassic Park* had earned the largest opening weekend ever for a movie up until that point. A Disney version of the Jurassic Park premise held a great deal of appeal.

As well as both of these premises tested, however, a clear winner emerged during the polling phase of Animal Kingdom. Disney pitched potential guests on the idea of a fantasy land predicated on the same fables that Walt had used as the backbone of some of his most famous films. Even with only a few details offered, pollsters lapped up the concept of what Disney was internally referencing as Beastly Kingdom. Their market research suggested that while all of their ideas for Animal Kingdom were good ones, Beastly Kingdom would sell the most tickets. Why would Disney choose not to build something that they knew would sell well? We'll get to that in a bit. For now, let's discuss the blueprints for the mystical realm that Imagineers never got to build.

Whether you love Disney's heroes or the villains that oppose them so aggressively, you understand the appeal of the bad guy. Even in the mid-1990s, Disney had already understood just how popular their rogues gallery was. Selling merchandise of the mortal enemies of Snow White and Cinderella doubled the revenue potential for the brands. More importantly, it established the thin line between love and hate, the gray area that Disney embraced.

As they plotted Animal Kingdom, some enterprising planner had a historic epiphany. Disney could build an entire themed land that pitted good against evil. Patrons wandering this land of mythical creatures would explore two regions. The realm of virtue and goodness would feature a unicorn as its centerpiece. The realm of dark malevolence would have the grander showpiece. A dragon would lord above the ruins of a castle. Perched atop its highest tower, his baleful glare would dare any challenger to approach the crumbling façade.

The twin paths were classic Disney to the core. They would provide theme park tourists with the ability to choose their own adventures. Anyone seeking serenity and light could walk the path of the unicorn. A thrill seeker could eschew kindness, daring to trespass in the place where angels fear to tread. The premise of Beastly Kingdom was epic in scope, thrilling to contemplate, and insanely appealing even to the most casual of Disney guests. It would have catered to a much more populist crowd than any other themed land at Walt Disney World, save for the ones at Magic Kingdom.

Unless you are a diehard *Fantasia* or It's a Small World fan, you likely wouldn't have preferred the "happy" side of Beastly Kingdom to its dark companion sector. The realm of good mythical creatures didn't have the sex appeal of evil for reasons that will become abundantly clear in a moment. That doesn't lessen the brilliance of the premise, though.

Let's call this side of the themed land the Good Place. Animal Kingdom guests wandering the path of light would discover a peaceful set of inhabitants living in an impossibly bright and colorful habitat. Gardens and ponds were a constant presence in the illustrations and early renderings of the Good Place.

The architecture intended to pay tribute to the classic Greek myths of yore in both subtle and overt fashion. Subtly, the buildings would honor the Greek design traditions, with Doric, Ionic, and Corinthian columns. Guests would feel like Odysseus as they encountered fantastic beasts in structures straight out of Homer's epic poems. As was the case in Homer's *Odyssey*, water was planned as a constant presence in the Good Place. Gentle streams and ponds would provide soothing reassurance in this utopia.

While a few minutes of peace and calm at Animal Kingdom sounds blissful enough on its own, the Good Place would also offer a pair of attractions. The easier to understand was Fantasia Gardens. Drawn as a loving tribute to and recreation of *Fantasia*, this ride would have the same structure as It's a Small World. A little boat ride would transport visitors to this idyllic land into the movie *Fantasia*.

Specifically, guests would visit the "Dance of the Hours" and "Pastoral Symphony" segments of *Fantasia*. As a Disney fan, your synapses should already be firing with imagery of dancing ostriches and hippos, elephants blowing bubbles, and female centaurs frolicking in the water. Understandably, music was planned as an integral part of the ride experience. Your musical vessel would play Ponchielli and Beethoven as you journey through the whimsical land of mythical fauns and cupids.

In recent years, Disney has mastered the art of re-creating the best scenes from their most popular movies. The concept of a boat ride through *Fantasia*, with accompanying visuals akin to The Little Mermaid: Ariel's Underwater Adventure, is bittersweet to contemplate. It would clearly stand out as one of the best and most family-friendly rides not just at Animal Kingdom but on the entire Walt Disney World campus. The fact that it doesn't exist is heartbreaking to any fan of the masterworks of Walt Disney the animator and filmmaker. And I'd also point out that one more dark ride that gives Animal Kingdom visitors a few moments off their feet and out of the sun...well, you know. Fantasia Gardens would have been a tremendous asset to Animal Kingdom for a multitude of reasons.

The other proposed signature attraction at the Good Place is a bit trickier to conceptualize. Entitled Quest of the Unicorn, it was projected as a maze environment that would appeal to fans of questing, in particular. Guests of the hedge maze would walk through explorable sections, each of which featured puzzles to solve. Yes, you read that correctly. This wasn't a ride per se, but instead a walking expedition. Many of the times Disney has plotted such attractions, they've wound up becoming rides instead, with Haunted Mansion as the most famous example.

The novelty of Quest of the Unicorn required guests to explore, comparable to Kim Possible World Showcase Adventure and Agent P's World Showcase Adventure at Epcot, although the Beastly Kingdom version would have been grander in scale.

While trying to solve the various puzzles, guests would interact with mythological creatures, all of them benevolent. These animals would guide their new friends to the correct answers, thereby activating five golden icons and sending guests along to the next portion of the maze. Once a person had completed all five sections, they gained admittance to the hidden grotto where the gorgeous unicorn resided. Presumably, this would have been a photo op for the ages, and the look of joy on a kid's face when they saw the fabled unicorn for the first time would have been unforgettable for parent and child.

As great as the Good Place sounds, it pales in comparison to the other half of Beastly Kingdom, which wouldn't have shown friendliness to strangers. To the contrary, its purpose was to dissuade potential visitors from stepping foot in these forbidden lands. The first step would involve crossing a bridge guarded by a troll. Disney strategized that they could make Beastly Kingdom even more foreboding by ostensibly denying entry. Their goal was a first-person perception of a waking nightmare.

Once you've vanquished the troll, you step into a dystopian nightmare of fantasy horror. Imagine, if you will, a dark twist on the villages from *The Sword in the Stone*. These medieval constructs have the look and feel of any basic town, save for one key difference. The center of town includes the same structures as Stonehenge, hinting at dark rituals performed against previous villagers. Other indicators of discord litter the ground. The broken swords and lances of would-be heroes reveal their fates. This settlement isn't welcoming to strangers.

As you enter the area surrounded by stones, your eye can't help but be drawn to the most impressive building in town. Formerly the castle of a beloved ruler, it's since fallen into a permanent state of disrepair. The castle stones look like they could collapse at any moment, and you wonder for an instant why no one has bothered with their upkeep. Then, your eye continues farther up the ramparts, and you suddenly have your answer.

A fearsome dragon towers above the castle. Its toothsome smile is readily apparent, even from the village below. Its breathing causes puffs of smoke to cloud the air, poisoning everything nearby. The posture of the most fantastic beast at Beastly Kingdom tells the story. You're in in this wyrm's domain, and the evidence surrounding you suggests that it slays all intruders.

Suddenly, you hear voices, whispers from above. You wonder for a moment if the sight of the dragon has driven you insane, but the sibilant murmuring continues. You look up and notice hundreds of eyes upon you. The dilapidated castle still hosts numerous residents. Mischievous bats live just beneath the dragon's current perch.

These bats have spent years plotting and strategizing the perfect crime. They seek to lure the dragon out of his lair long enough to steal the priceless treasures stored within. When they take notice of you, the first new visitor in ages, they hatch a ploy. The bats use an odd combination of mockery and humor to cajole you into stepping inside the castle. You know it's an act of sheer madness, yet you feel compelled to enter. The mysteries held within are too tantalizing. In this moment, you cast aside all your fear and doubt and step inside the Dragon Tower, ready to meet your fate.

As you can tell from the above description, the sole attraction on the dark side of Beastly Kingdom was epic in scope. Imagineers had the wildest ambitions for the linchpin ride at Animal Kingdom. They fully expected the Dragon Tower to stand as the primary monument not just in this themed land but for the entire park. During the blue sky phase of planning, they unleashed their creativity in plotting a ride epic in scope and truly badass in nature.

Make no mistake on this point. Imagineers designed Dragon Tower as a roller coaster on a par with California Screamin' or Rock 'n' Roller Coaster Starring Aerosmith, both of which would debut soon after Animal Kingdom. At the time, Disney was sensitive to the criticism that is parks lacked true thrill rides. The signature attraction at Beastly Kingdom would emphatically end those arguments.

The expected version of Dragon Tower would include your interactions with the sardonic bats. They'd goad you into boarding the ride and heading toward the proverbial dragon's lair. You can figure out what would happen next. The dragon would stir from its slumber, take note of its unwelcome guests, and proceed to chase you through the various rooms of the desolate castle. A few fireballs straight from the dragon's maw would singe you a bit as you tried to escape. Eventually, you'd narrowly survive your encounter, leaving you and the bats in the belfry to split the spoils of your heist.

Basically, depending on your favored medium, Dragon Tower would place you in the role of Dirk the Daring from Dragon's Lair or Bilbo and his team of dwarves in *The Hobbit: The Desolation of Smaug*. Based on the drawings I've seen, I say in complete sincerity that this is the best dark ride concept in the history of Disney…and I'm saying that while acknowledging that Disney is the originator and master of the dark ride. It would have been magnificent. While the Good Place would have had its supporters, the dark side of Beastly Kingdom would have become the biggest draw in the park and possibly anywhere at Walt Disney World outside of Magic Kingdom.

Why didn't any of this happen? Well, mistakes were made…

Despite spectacular net income in 1996 and 1997, Michael Eisner felt the need for caution in building his latest themed park. Many of the people involved with the design of Animal Kingdom worried that fictional characters would distract from the zoo premise, even as they corrected anyone who had the audacity to describe it as a zoo, a huge no-no in Disney's staff rooms. They felt the same way about dinosaurs, but they recognized they weren't going to win that battle for a reason we'll discuss in a moment.

Fictional creatures became a subject of a vast schism within the walls of Disney. It was a rare instance where the artists and the bean counters lined up on a subject.

Disney's most creative employees relished the idea of letting their imaginations run wild, re-creating some of their favorite stories in a themed land that existed for precisely that reason. It was the first chance in 40 years to walk in the same footsteps as Walt Disney, who created Disneyland as a way to bring people's dreams to life.

For the bean counters, the situation was more basic. Disney believes strongly in anchor products at theme parks. The reason why so many landmarks exist at each of their gates is hidden in plain sight. Cinderella Castle, Space Mountain, and Spaceship Earth all build awareness. They're static monuments that have a second function as permanent billboards, warmly inviting guests inside. You may see a giant golf ball when you approach Epcot, but what you're really noticing is a sublime marketing tactic.

Beastly Kingdom would have this hook. The dragon lording over his domain would menace onlookers, accidentally adding mystery and intrigue to the lair it guarded. Guests would also embrace the challenge of overcoming the five great feats in order to gain an audience with the unicorn. Even ignoring both of these tactics, a more basic aspect appealed to the number crunchers. The data suggested that Beastly Kingdom was the ultimate draw at Animal Kingdom, not the zoo element. Most major cities have zoos, after all. The pervasive belief before the Disney version became a reality was that they couldn't do much better than the ones already in existence. We know now that this was a short-sighted philosophy, but it explains the polling data. Beastly Kingdom promised something new and different, and that made it a draw.

When Eisner had to choose how to disburse his resources, he weighed all the options and promptly made a *terrible* decision. He sided with those who believed Animal Kingdom should highlight its animals, an understandable philosophy. Where history would bemoan his judgment was in the second call.

When facing two options and understanding that he could only afford one, Eisner chose to pay tribute to dinosaurs with DinoLand U.S.A. It was the Sam Bowie over Michael Jordan moment in theme park history. Disney spent the bare minimum in bringing DinoLand U.S.A. to life, and even the most casual of theme park tourists could tell. From day one, it felt like a hastily thrown together attempt to boost the total number of attractions and distractions at Animal Kingdom rather than a viable themed land.

Meanwhile, Eisner promised all his park planners that Beastly Kingdom would become an integral part of phase two at the park. Going back to the original Disneyland, a quick series of updates was a staple of the company. While they hadn't excelled as much in Eisner's tenure, he planned for Animal Kingdom to change the perception of his leadership and vision. Beastly Kingdom was so critical for the quick expansion of Animal Kingdom that he included it in his introductory speech. The silhouette of a dragon is also part of the park logo to this day. Everyone involved was that confident of its inevitable arrival.

Why didn't Beastly Kingdom become a part of phase two of Animal Kingdom? Let's now have an odd discussion about park attendance and its shocking impact on the fate of Beastly Kingdom, starting with a simple question: why did Disney build Animal Kingdom? The answer isn't altruism. Yes, Walt Disney himself was a champion of environmental causes and animal conservation. Yes, even the early days of Disney television series featured interactions with nature, a premise they've honored with their recent line of Disneynature movie releases. But the corporation's love of planet Earth and all of its inhabitants only goes so far. It's money that matters.

The core concept to Animal Kingdom was that it would secure the short-term future of Walt Disney World after the hiccup at Disney-MGM Studios. Everyone involved from Eisner down to the Imagineers to loyal cast members believed that a fifth park was only a matter of time. Still, Animal Kingdom stood above the other 1990s structures as the most important project. Its overriding purpose was to boost attendance at Walt Disney World.

Even the harshest skeptics of this project never questioned its appeal. Analysts and fans alike universally acknowledged that their fears about Animal Kingdom involved functionality, not drawing power. Building and sustaining innumerable ecosystems in plain sight of the viewing public seemed like a daunting if not impossible situation for Disney. Everyone agreed that if they pulled it off, they'd reap the financial rewards.

Oops.

A concept called cannibalism exists in the theme park world. It's not quite as Donner Party as it sounds. Disney execs describe one of their worst traffic problems by using this particular term. To them, cannibalism occurs when the creation of some new attraction or exhibition pulls attendance away from other, similar venues.

As an example, when Disney builds a new restaurant, they don't want to siphon attendance away from existing restaurants. Instead, their goal is to expand the overall appeal to eating at a Disney theme park rather than bringing your own food or leaving the park to eat somewhere else. Disney is now paying additional costs for constructing and staffing a new business on site. They need to counter-balance those expenses by earning more revenue thanks to customer patronage.

The concept of cannibalism drives a surprising number of decisions at the various Disney theme parks. Shows and fireworks in particular are worrisome since these activities require a lot of a guest's free time. Disney has to structure the displays in a way that attendees will remember to keep their wallets open. Otherwise, by offering such wonderful distractions, Disney decreases its own bottom line each day.

The ultimate cautionary tale of cannibalism at Walt Disney World is, perhaps fittingly, Animal Kingdom. After spending almost a billion dollars to build and staff their fourth gate, Disney execs were euphoric about its early reviews. Virtually all of them championed the park as a hallmark achievement in theme park design and execution. Of particular note to critics was that Disney somehow introduced thousands of animals to the natural habitat of central Orlando without endangering them or the strangers suddenly interacting with them on a daily basis. And a lot of folks, myself included, were shocked that Animal Kingdom didn't have that zoo/county fair smell. In short, Animal Kingdom's debut was the best-case scenario for Disney, a much-needed win after a few years of theme park disappointments.

Before Michael Eisner could take his victory lap, the attendance numbers started to pour in. They were...surprising. Yes, attendance was solid at Animal Kingdom. In its first year, which was really just the last eight months of 1998, six million guests attended. That was a larger total than Universal Studios Hollywood managed for the entire year, and it crushed a couple of other popular local theme parks, SeaWorld Orlando and Busch Gardens Tampa Bay, too. That was the good news.

The bad news for Disney and Eisner was that attendance at all three other gates at Walt Disney World dropped in 1998. And then they dropped again in 1999. It's true. In 1997, Magic Kingdom counted 17 million visits from guests. The following year, that total dropped a hefty eight percent to 15.6 million and then fell another three percent to 15.2 million in 1999. Epcot started with 11.9 million in 1997 but finished 1999 with only 10.1 million. Disney-MGM Studios also fell from 10.5 million to 8.7 million.

Take a moment to think about these statistics from the Disney perspective. They had just invested a billion dollars to earn a traffic spike from loyal theme park tourists. Instead, something wildly unexpected had happened. Animal Kingdom had cannibalized the other three gates in a historically unprecedented manner.

Yes, the fourth gate garnered 14.6 million in attendance in 1998 and 1999. At the same time, the other three gates lost 9.1 million tourists over a two-year period, with the losses being measured against 1997 park traffic. In other words, MGM fell by a million in 1998 from 1997, and its 1999 total was down an almost incomprehensible 1.8 million from 1997 totals. Disney's "net gain" was 5.5 million park visits over two years, a fraction of what they'd projected. The worst news of all for Disney is that every lost customer meant lost revenue in the form of meals, merchandise, snacks, hotel stays, and incidentals. They had anticipated a packed Walt Disney World in the wake of their new Animal Kingdom offering. Instead, the fourth gate had Idi Amin'd the other locations, offering very little tangible financial benefit.

What's the consequence of this revenue shortfall? For a multi-national conglomerate like Disney, the areas of impact are myriad. The Parks and Resorts division understandably felt it the most since their portion of the ledger was the one with the most troubling numbers. Disney not only failed to capitalize on the addition of a fourth gate, but found itself struggling to pay for the venture due to shaky attendance figures across the entire Walt Disney World campus.

You don't need a degree in economics to see where this is going.

The underlying strategy of Animal Kingdom was to entice theme park tourists to expand their travel plans. Guests would now need to stay an extra day at Walt Disney World to see and do everything. Animal Kingdom was for all intents and purposes a strategy Disney execs employed to increase guests' vacations from four days to five or from six days to seven. This may not seem like a huge deal to you, but it's as much as a 25 percent increase in revenue for a theme park owner. Disney was *counting* on that income, first to pay for the cost of construction and then later to become a source of evergreen revenue. The traffic shortfall was a setback with far-reaching consequences felt for years in Orlando.

With Animal Kingdom having failed in its primary task, improving Disney's bottom line in the Parks and Resorts division, reality hit home for most of its champions. Even the most ardent supporters of the concept appreciated that the CEO and board of directors wouldn't look at Animal Kingdom as worthy of further expenditures, at least not in the short term. Any hope of a massive phase two expansion at the park fell by the wayside.

In between the lead-up to Animal Kingdom's opening and the planned phase two, the most fascinating part of the story occurred. It's an integral part of theme park folklore, something that's had a shockingly significant impact on the recent history of the industry.

When Eisner and his team at Disney recognized that their Animal Kingdom project was too expensive, some bloodletting occurred. He released some Imagineers from their contracts, which is a polite way of saying that he fired a bunch of employees. He didn't like how combative his underlings became over the news that Beastly Kingdom wasn't moving forward, at least for a time. And he also didn't like that the cost of his park had ballooned to almost a billion dollars. For better or for worse, the constant claim about Michael Eisner was that he cared more about money than people. Fate punished him for this character trait, and in a very strange way.

As park planners for Universal Studios Florida plotted an expansion of their park, they suddenly enjoyed a windfall of employment opportunities. Disney Imagineers, some of them with decades of experience, suddenly entered the open market as free agents. It was an unprecedented opportunity for Universal, as Disney had historically safeguarded their Imagineers from any change of competitor poaching. Now, Eisner had nonchalantly dumped them. To Universal execs, this was found money. They wanted to build a new gate in Orlando, and they suddenly had the opportunity to hire some of the best theme park creators on the planet.

You probably know the history from here. These Imagineers fearlessly, perhaps shamelessly, borrowed from the plans for Beastly Kingdom. They took the concept of Dragon Tower and evolved it in Dueling Dragons, a ride now known as Dragon Challenge. When Universal re-themed Islands of Adventure as the Wizarding World of Harry Potter, it was one of their updates. That attraction owes its history to Beastly Kingdom.

Similarly, the second phase of Harry Potter at Universal, the Gringotts Wizarding Bank, has an unforgettable attention-grabbing façade: a fire-breathing dragon sits atop a tower. Draw your own conclusions about the origin of this premise.

Universal has unmistakably co-opted some of the ideas from Beastly Kingdom. In the process, they've leveraged these concepts into a popular pair of themed areas that have led to a resurgence of the second most popular theme park in Orlando. And former Imagineers that Eisner fired are the ones leading the charge.

As Animal Kingdom approached its target opening date, Earth Day of 1998, one more bit of bad news hit Eisner's desk. Park planners had serious concerns about the volume of park attendees relative to the number of attractions and areas available. They felt the pinch of not having the three spectacular attractions from Beastly Kingdom as well as the expected walking area of that themed land.

After a brainstorming session, Disney's upper management hastily threw together a permanent summer camp known as Camp Minnie-Mickey. It originally featured a pair of live shows, only one of which has stood the test of time. The failed endeavor was Pocahontas and Her Forest Friends. It lasted a decade, but claimed sparse attendance in its final years. The more popular stage show remains today: Festival of the Lion King, the best thing to come out of the failed attempt at building Beastly Kingdom. Other than that show, Camp Minnie-Mickey was largely a glorified time killer with some character meeting spots. Disney mercifully shut it down in 2014. And the reason for that is the second postscript.

In 2011, Disney shocked theme park observers, movie analysts, and basically everyone in the business world when they announced a new expansion for Animal Kingdom. Whether you want to call it Phase Two or something else, Pandora: The World of Avatar debuted in 2017. It exists in the same space that Eisner and a team of optimistic, creative Imagineers had once slotted for Beastly Kingdom. It is unquestionably better than Camp Minnie-Mickey, but Disney historians will always lament that it takes the place of Beastly Kingdom, the best Disney themed land that never got made.

CHAPTER NINE
Pleasure Island

Pleasure Island sounds like a place where clothing is optional, doesn't it? At least it would if you weren't a Disney fan who already knew what it was. For this reason, the idea of Pleasure Island at Walt Disney World always sounded incongruous to outsiders. Pleasure Island is a Las Vegas-styled name deployed at a place decidedly not targeting adults, at least not primarily.

Conversely, Lake Buena Vista Shopping Village sounds like a happy retirement community where retirees invite their grandchildren to spend the holidays. Therein lies the logic that Disney execs used in trying to unearth new ways to improve their bottom line. They wanted to maximize their revenue potential during the early 1970s, the trying period that came soon after Walt Disney's death. The best way to do that was to build an entirely new development, one that they could market to every adult on the planet. This entertainment district was a transparent attempt to expand the target audience for Walt Disney World.

Did the plan work? The answer depends on who you ask. The truth of the matter is that the portion of Walt Disney World that we now know as Disney Springs wasn't even part of the core plan for Walt Disney's city of tomorrow. It was a previously planned shopping center outside the experimental community. The historical records for Lake Buena Vista suggest that it was to stand as the highlight of a complimentary town close to EPCOT, but not for residents of the utopia. It was always intended as a showcase for something better, even before the Florida Project evolved into something different.

What's strange about Disney Springs is how many iterations of its existence have produced the same outcome. In the early days, it was one of the most popular commercial shopping areas in the south. Then, an imitator became the preferred choice in Orlando, Florida. In the wake of this turn of events, The Walt Disney Company tried to get smart. They imitated the imitators.

Of course, the Disney style required a theme, and that theme was…Merriweather Adam Pleasure. You may not know the name, but you know the outcome. Disney debuted a new shopping mecca that they named Pleasure Island, an inaccurate title from the corporate perspective. This free portion of Walt Disney World sounded great in theory. In execution, well, you know.

Disney strategists tried everything to avoid negative headlines and ugly financial statements with their entertainment district in downtown Orlando. Their attempts were largely unsuccessful, and that's why theme park tourists now hang out at Disney Springs. It stands on the shoulders of Pleasure Island and all that came before it. Let's take this opportunity to learn the history of Disney Springs or, as people knew it in 1975, Lake Buena Vista Shopping Village.

We'll discuss all the hopes and ambitions Walt imbued in the premise, how his successors attempted to honor his intentions in building a community fixated on shopping, the competitor that forced Disney to take their development more seriously, the miscalculations involving a man named Merriweather, the underlying theming of Disney Springs, and its future. Yes, it's a lot, but when you're finished with this chapter, you'll have newfound respect for that place with all the great shopping and food. Disney Springs is a triumph of corporate stubbornness. Read on to learn the how and why of it.

Nothing screams utopian society like shameless commercialism. Yes, that's admittedly a bit glib. I struggle not to laugh when considering the seemingly contradictory parts of the Experimental Prototype Community of Tomorrow dream. Walt Disney wanted to build a better world predicated on a world-class working class. The residents of EPCOT would sing for their supper, so to speak. Everyone who lived there had to hold down a job, and Walt projected those employment opportunities to involve cutting edge technological innovations. The blueprints for EPCOT homes even included suggestions that workers would return home to discover that they'd received upgraded appliances while at work.

The worker-bot society benevolently rewarded by its leaders sounds lovely in theory. In execution, a problem exists. How would Disney pay for all this? The early expectations for EPCOT involved other corporations. They would pay for the right to visit EPCOT and participate in joint projects together. Effectively, the honor of working hand in hand with Imagineers would prove inviting enough for corporate America to throw tons of money at the Disney utopian community. Thanks to many career successes and coming on the heels of Disney's triumphs at the 1964 World's Fair, the remarkable innovator and company founder felt confident about his strategy.

The second ingredient in Walt's mercantile concoction would happen at the on-site World's Fair. Disney felt so euphoric about the proceedings at the New York event that he embarked on a trajectory that would ultimately lead to the World Showcase at Epcot. That wasn't the original idea, though. When planning his utopia, Disney shrewdly deduced that visitors would come from other parts of the East Coast as well as around the world to visit the new version of Disneyland.

This amusement park would prove integral to the revenue stream of the Walt Disney World campus in time, but Disney always viewed it as a rainmaker. He also wanted to augment it with a living World's Fair, an exhibition of global culture, heritage, and foods. Long before EPCOT Center opened its doors for the first time, the man who bought the land and had the plan already knew how the commerce would work. This World Showcase would sell international merchandise (at a healthy markup, of course) and provide foods of the world, including local favorites like southern cuisine and general American cuisine.

These revenue streams would also help to pay for the community as it thrived. Having spent so much of his life paying to keep the lights on at his various buildings, theme parks, and movie studios, Walt Disney understood as well as anyone in the 20th century the importance of evergreen cash flow. He even doubled down on the premise of monetizing visitation at his central Florida utopia.

While studying the available land in the greater Orlando area, he noted that one of the overseers of the Florida Project had crossed out an area of land. The project manager had spent tons of money determining the viability of potential land acquisitions in the region. He'd discovered that this particular parcel of land would require lots of jumping through hoops to acquire. In other words, it would involve a time delay.

Time was of the essence to Disney since they were employing subterfuge to purchase the swampland. Ostensibly, the acreage had zero redeeming qualities. It should have had limited property value. If Walt Disney wanted the land, however, its value was the equivalent of Boardwalk and Park Place. The employees in charge of finding acreage for him to buy crossed off anything that would require a lot of time and effort. They needed people willing to sign on the dotted line immediately in exchange for some quick cash. Had they known it was Disney money, the prices would have skyrocketed.

All of this is germane to the Disney Springs discussion due to an odd bit of trivia. When Walt looked at a map of central Florida, he noticed the very region that legal counsel Bob Foster had crossed out. Where Foster saw a subdivision of ownership interests and a protracted set of negotiations, Disney saw convenience and maybe a few dollar signs. The troublesome parcel of land resided near the interchange of Highway 535 and Interstate 4. It was an easily accessible spot, a priority to Disney, a man who learned to exalt logistics after a few miscalculations at Disneyland.

Foster would later lament that Disney's obsession with that piece of land delayed the entire Florida Project by a full year. It was an instance where everyone involved made the correct assessment. The overseers and legal counsel correctly evaluated the complexity of the land acquisition. They were right to state that it might not be worth the aggravation. Meanwhile, Disney proved prescient in recognizing that the convenience of the locale would lead to heavy traffic over time.

When elderly Floridians discuss the development of the Walt Disney World campus, they're inclined to highlight the constructions of Magic Kingdom and EPCOT Center. The former theme park gate fundamentally changed the perception of the state while the latter, well, some lamentations exist about what was promised versus what was delivered.

What doesn't get mentioned enough is the construction project that happened in the middle of the two events. Magic Kingdom opened in 1971, and EPCOT

Center followed more than a decade later, in 1982. Disney Imagineers weren't sitting on their hands during this gap. To the contrary, many of the most important projects in theme park history owe their origins to the early 1970s, Space Mountain being the most important one.

The problem that Disney perennially faced as a company then and still does today is paying for new attractions and features. During the early days of Magic Kingdom, park planners had to kill their darlings on a fairly regular basis, picking new rides at the expense of others. As we learned with the arrival of Shanghai Disneyland in 2016, cost-cutting still goes part and parcel with progress.

One of the skills Disney developed over time has its origins at the Walt Disney World campus, midway between the arrivals of Magic Kingdom and EPCOT Center. In 1975, the company introduced Lake Buena Vista Shopping Village. It seemed like a rather innocent shopping center at the time, and there was no reason for customers to appreciate the bittersweet nature of its existence. This strip of stores and restaurants was the culmination of an idea that Walt had carried with him more than a decade. The land he'd once seen on a map of central Florida had finally become the shopping area/breadwinner that he'd forecasted. But the journey to reach this point was lengthy and meandering.

Independent of how or why guests visited the Walt Disney World campus, they'd still share the same needs. Those included food, shopping, and recreation. While strategists spun their wheels trying to find a palpable version of EPCOT, they could do *something* with part of the land. Critically, this entertainment section of the Disney land could build an evergreen revenue stream for the company, one they sorely needed in the early 1970s.

At first, Disney employees weren't trying to sell merchandise to theme park tourists. The new establishment had a different target audience. Disney strategists recognized that the quickest way to infiltrate the community and become trusted new neighbors was by providing services to the locals. Of course, Disney planned to ship "the locals" into Florida rather than sell to the ones there.

Overseers plotted the Lake Buena Vista community as a place where Disney fans and employees would live, assuming that they'd later move to EPCOT to enjoy the utopia. They'd presumably spend all their money on Disney goods and services while living there, and they'd have to work for the company in some field to remain. That was one of Walt's stated policies about the new community. The Buena Vista Shopping Village was an insidious way for the company to "pay" its residents with salaries that they'd quickly reinvest in the local economy, all of which was controlled by Disney.

This strategy fell by the wayside as the reality of Walt Disney's absence began to hit home. His utopia would never come to fruition, and Disney employees wouldn't reinvest their salaries on Disney goods and services. Wait a second. That part kind of came true. Anyway, the point is that the plans for this region of land were decidedly corporate.

The change came when Disney modified the target audience. At the start of 1972, they settled on the idea of selling one- and two-bedroom villa homes to other corporations. Internally, Disney described their new community as the Lake Buena Vista Club, a themed community sharing the intimacy of Disneyland, only in a much larger overall region, Walt Disney World. The overseers felt that demand for Walt Disney World would cause a new kind of land rush. Workers outside of Disney would want to visit so much that clever corporations could incentivize them. The best employees during a given quarter/year would earn the right to stay at a villa on the Walt Disney World campus.

The premise was misleading. The path from the villas to the fun part of the area, Magic Kingdom, was many years away from becoming convenient. Nobody living there would feel close to the park even though they technically were. Still, Disney's marketing team went above and beyond in selling this real estate development. They described it as "the good life," promising world-class golf, dining excursions so elitist that wine captains handled the pairings, and the greatest social scene this side of Vegas.

The whole thing was over the top, but the underlying premise was simple. Before Disney could build EPCOT, they had to, you know, build a city. The running joke about Lake Buena Vista was that it was the experimental prototype of EPCOT. Disney anticipated that corporations would buy any and all condominium-style villas that they offered. That would take care of many financial constraints the company faced at the time. Their expectations proved misguided, but the villas Disney created would eventually sell to non-corporate travelers seeking a second home in Florida. In a way, the members of the Lake Buena Vista Club were the forerunners to the Disney Vacation Club.

Equating the Lake Buena Vista Shopping Village of the 1970s to the Disney Springs of today is like comparing a silent movie from 1908 to *Inception*. We've come a long way, baby. That doesn't diminish the achievement of the initial iteration of the premise, though.

From the beginning, this shopping complex showed entrepreneurial promise. Impeccably positioned near State Road 535, the development fittingly rested on Preview Boulevard (near the legendary Preview Center) just beside Lake Buena Vista, a watery parcel of land that Disney owned.

In fact, Disney had such power over the Florida government thanks to the Reedy Creek Improvement

District that they could determine the location of the water. The autonomy they received in the agreement was so wide-ranging that they had the ability to name and divide the acreage as they saw fit. Disney actually carved up some of Bay Lake and some previously unincorporated land to build Buena Vista Lake, the name reflecting the street address of The Walt Disney Company in Burbank, California.

Once corporate execs understood who was going to live in the fledgling community and visit the Lake Buena Vista Shopping Village, they knew how to build it. Their consumers weren't the corporate power brokers Disney had anticipated. Instead, they were theme park tourists like you and me, only the 1970s equivalent, guests who were also interested in a second home. It was a "town" of itinerant visitors, meaning that this entertainment district would always feature guests flush with money. Everyone loosens their wallet while traveling. Strategists could do the math from there.

During its opening in March of 1975, the new shopping village featured upscale entertainment options. Their design was to sell the idea of Walt Disney World as the good life, a patrician escape from society. Some of the earliest plans actually referred to the main throughway as Royale Circle, with one of the original ideas honored by loyal Disney execs. The *Empress Lilly*, named after Walt Disney's wife, showed on the blueprints as an elegant paddleboat. It would offer the finest cuisine in the village.

This particular concept has stood the test of time. You know it now as Paddlefish Restaurant, although its more recognizable recent name was Fulton's Crab House. When you eat here, you are dining precisely where Walt intended when he drew up the plans for the Walt Disney World shopping district. That's reason enough to book an advance dining reservation, right?

Other offerings on day one at Lake Buena Vista Shopping Village included an upscale home décor store named Bath Parlour, an early imitator of Williams and Sonoma known as Gourmet Pantry, and a party supply store with the glorious name of 2 R's Read'n & Rite'n. The village also offered a showplace known as Captain's Tower, a seasonal promotions office and a great place to take kids to give them a bird's eye view of the area. Perhaps the most Disney-fied store at the shopping district was It's a Small World After All, a venue that sold…you guessed it, children's clothing. Forgive them their puns. It was the 1970s, my friends.

The other luxury store at the village was the most telling. Named Posh Pets, it catered to wealthy clientele in need of a new furry family friend. They also offered grooming and other luxury pet caretaking on site. It was telling because Disney was fostering the perception that their visitors at Lake Buena Vista would develop an entire lifestyle there, up to and including pet ownership. Disney did everything they could to make guests feel like they were home, albeit at a huge price markup. Let's be clear that anybody who bought man's best friend here paid a stiff fee for Fido. That was the recurring theme of the first iteration of Disney Springs. History truly does repeat itself.

Expansion over the years is where the village went wrong. Starting in 1977, campus planners increased the business offerings in hodge-podge fashion. They also changed the name to Walt Disney World Village to reinforce its appeal to consumers. Expansion was inevitable at this point. No reason or rhyme existed in adding more stores, though. It was simply a matter of earning more revenue. That caused problems, the most pressing of which involved the basic design of the place.

During the early days of Walt Disney World planning, Imagineers agonized over every tiny detail of construction. They sweated over minute theming details such as building everything in a series of triple circles that look vaguely like mouse ears. With Lake Buena Vista Shopping Village and the surrounding community, they implemented hexagons instead. And the thing about the hexagon is that it can only have six points to maintain its structural integrity. Anything more than that turns it into, well, a heptagon. In this way, Disney's grand vision for the entertainment district was remarkably fragile.

At its genesis, the village accentuated the vacation kingdom theme of Walt Disney World in its entirety. An entire community surrounded a man-made lake at the center of town. The only expected vehicles were WEDway PeopleMovers, which explains why generations of Disney Springs visitors have lamented the parking situation. It was a conscious choice on the original blueprints. Somebody from the late 1960s deserves a punch in the face for that.

Anyway, part of this village-wide philosophy involved "commercial naturalism," a concept that has long since fallen by the wayside. At the time, it was a belief that everything should look organic and lived in. Shrubs, trees, and other vegetation seamlessly integrated with the shops and restaurants. Aged brick lined the walls, occasionally interrupted by rustic iron gates and stained-glass windows and lanterns. All these features were breathtaking to behold on day one.

The patterns were also shockingly easy to disrupt. A single reckless addition could ruin the illusion of theming that Imagineers had planned so meticulously. After the first year of its existence, the village entered into a battle of art versus commerce and, as usual, art got its butt kicked. The region haphazardly introduced new businesses into the village, accidentally ruining its timeless symmetry in the process. Once the 1980s arrived, any remaining hexagons were difficult to spot in a sea of interchangeable shops. In its constant fight for revenue, Disney lost the theme of one of its most beautiful creations, the Lake Buena Vista Shopping Village.

As Disney fell by the wayside, an unlikely threat emerged. And they were largely responsible for it. When Walt purchased his land and announced the Florida Project, Orlando had a population of 250,000. That number had more than tripled as the end of the 1980s approached. The city would surpass one million residents in the early 1990s. A lot of the appeal was living in a thriving metropolitan region that was home to Walt Disney World.

As the population increased, other entrepreneurs noted the growing opportunities in the city. Church Street Station was arguably the most famous. It was a series of nightclubs hosted in the unlikely setting of a train station. Guests would literally cross the tracks as they passed the nights away, barhopping in an impressively enterprising setting. For a single cover charge, they'd have free reign of every night club there. And Disney viewed basically every dollar spent at Church Street Station as money taken from their ledger sheet. It was their idea, so they were furious to watch somebody else monetize the greater Orlando community by employing their strategy. Better.

Over time, strategists appreciated that one of the reasons Church Street Station succeeded at the village's expense was its target audience. Disney had explicitly chosen to emphasize a luxury lifestyle. They wanted to sell to the wealthiest clientele, a rejoinder that sounds familiar to this day. Church Street Station was the populist response, a welcoming environment that would happily take money from people of all income levels. Disney had unintentionally priced out some of their potential customers.

Appreciating the problem, corporate executives finally embarked on a new strategy as the 1990s approached. They decided to reboot the village as a vacation destination with a thriving nightlife. Out was the proverbial Grey Poupon. The new spot was all about mustard instead. And that was the genesis of Pleasure Island, the locale that we now know as The Landing at Disney Springs. In the years leading up to the current iteration of the village, however, Pleasure Island was very much its own thing, and a weird little thing at that.

Does the name Merriweather Adam Pleasure mean anything to you? Hopefully the answer is no because even Disney superfans struggle with this one. Mr. Pleasure (or MAP, if we use his initials) is the fictional founder of Pleasure Island. Disney created this character in order to theme the new version of the entertainment district. They no longer fixated on aesthetics at the expense of actual story, which was in reality a step back, something they wouldn't recognize for too long.

While the village had a style all its own, Pleasure's new "island" emphasized shopping with a bit of back story. And the weird back story was that in 1911, MAP bought himself a new plaything, an island. Bored during downtime from one of his many adventures as a member of the Society of Explorers and Adventurers (S.E.A.), the itinerant explorer put down roots in his new home. And, like a good megalomaniac, he named it after himself. This tidbit was critical to the development of Pleasure Island since its signature restaurant was the Adventurers Club.

This nightclub was singularly unique in the history of Disney, and it's so revered by some that it's worthy of a standalone discussion at some point. Somehow a hybrid of Jungle Cruise narration, improvisational comedy, audio-animatronics, and Rick's Café Américain in Casablanca, the Adventurers Club was the place to go for a great time at Pleasure Island from 1989 until

Disney closed its doors in 2008. During the final few days of its existence, fans flocked to it similar to how attendance spiked at the Osborne Family Spectacle of Lights in 2016. They even held a couple of memorial tribute shows in the years after S.E.A. shuttered the Adventurers Club.

The theme of the nightclub tied into the overall theme of Pleasure Island, and that theme was fun. Of course, Imagineers thought it was Mr. Pleasure, who viewed the site as a place to relish in his "lifelong interest in the exotic, the experimental, and the unexplainable." When the would-be entrepreneur arrived on the island, he plotted to become rich by manufacturing sails. Only one problem existed. Sailing was a dying industry as the Wright Brothers had recently developed the power of flight. Pleasure should have gone bankrupt, but World War I saved him. Military navies needed sails, so a stupid idea became an insanely profitable one. Merriweather had cause for merriment.

The reason why new ownership was in place when the Adventurers Club opened was due to (hilarious) tragedy. Pleasure, the "Grand Funmeister," made the incorrect decision to explore Antarctica, a journey from which he never returned. His bumbling sons drove his estate into bankruptcy, and they lost Pleasure Island in 1955 when (the real) Hurricane Connie wiped out the (fictional) community, which is a dangerous theming for a Florida resort...but I digress.

With no money and a destroyed community, the sons of Pleasure (how is that not a band name?) gave up on their exotic real estate inheritance. It devolved from a hotbed of decadence to the proverbial deserted island until archaeologists discovered the abandoned lands in 1987. Like any good archaeologist would, they chose to build a shopping and entertainment district as a monument to the weirdness of Merriweather Adam Pleasure.

Disney theming is always the gold standard, but their decisions with Pleasure Island were universally daring. They wanted to add a sense of whimsy to the experience of visiting Walt Disney World. The former site of the village became a new kind of entertainment district, one that prioritized parallels between the fictional themes and the very real stores.

Disney even elevated the proceedings with a nightly fireworks show. They called it a celebration of New Year's Eve, which somehow occurred every single night. If true, that would mean that Pleasure Island was open for about 5,500 years. Alas, it only felt that way to employees. The forced celebration was so memorable that a Simpsons writer later recounted it in the "Itchy and Scratchy Land" episode. The applicable exchange is:

Marge Simpson: It must be wonderful to ring in the New Year over and over and over.

Cast Member: Please, kill me.

Innumerable former Pleasure Island employees are nodding their heads at this comment, but the festivities had their intended effect. Many theme park tourists grew passionate about the entertainment district. And people loved the Adventurers Club, arguably the most fun environment in the history of Lake Buena Vista Shopping Village/Pleasure/Island/Downtown Disney/Disney Springs.

The rest of the Merriweather Adam Pleasure theming wasn't quite so overt. Sure, anyone reading all the plaques and signs could add up the clues, but what Disney really wanted from their shopping center was money. They decided that the best way to get it was to tie everything together via story tales. Unfortunately, history repeated itself as Pleasure Island expanded and changed. No shopping area will stay the same over 20 years, of course, but any alteration to a themed area causes ripple effects.

A few people knew the history of Videopolis East, Fireworks Factory, and Merriweather's Market. When Pleasure Island swapped them out for 8Tracks, Motion, and Raglan Road Irish Pub, something was lost. Fireworks Factory was supposed to relay the ridiculous tale of MAP using a cigar to blow up a building (accidentally, of course). What that has to do with Irish cuisine is a mystery only Disney understands.

Videopolis East was perhaps the most egregious instance of adapting too much and caring too little about theming. In 1989, it started as a New Wave dance club, making it at least five years too late. Then, it switched to Cage, which was vaguely alternative with a bit of grunge thrown in. At least Disney had the timeline right there, adding it in 1990. Then, they switched 8Trax in 1994; it was themed as a Studio 54-ish dance club, with a blend of 1970s and 1980s music. With the benefit of hindsight, that would have been the perfect fit in *1989*. With three different themes in only five years, the place always struggled with a reputation as a cynical nightclub where Disney pandered to whatever they thought crowds wanted at the time.

Whenever Disney changed something, Merriweather Adam Pleasure died a little bit more. Of course, it's not that tragic since he never existed in the first place and also was kind of stupid to boot. A phantom moron as the primary theme of Pleasure Island speaks volumes about the flaws with the concept. Still, everyone who hung out at Pleasure Island generally had a great time. They were still at Walt Disney World, after all. And the company expressed satisfaction with the entertainment complex as long as guests continued to attend.

The change happened in the early 2000s. And it wasn't Disney's fault. The American economy took a turn and then the global economy grew sluggish, too. The prevailing opinion about Pleasure Island was that it was inferior to Universal CityWalk Orlando, an entertainment district a decade newer and with better anchor restaurants. Universal employed a different kind of theming, building anchor establishments for popular franchises like Margaritaville, Bubba Gump Shrimp Co., and Hard Rock Café. Despite the overwhelming charisma of Merriweather Adam Pleasure, many tourists stuck with the brands they knew rather than a flamboyant playboy adventurer.

By the time the housing market collapsed and the global economy dipped precipitously, Pleasure Island was doomed to a spot in the more frustrating annals of Disney history. Everyone agreed that the shopping center sold well and was a profoundly forward-thinking idea by Walt Disney. Its execution simply was lacking. The erratic nature of updates to the village caused it to lose thematic integrity. It became gangly and unattractive over time. Then, Pleasure Island mixed the basics of restaurants and shopping with the outrageously eccentric and weird. Disney recognized that they have fans of each aspect of the experience, but finding a person who enjoyed all three was rare. And the cost of upkeep on the Pleasure Island nightlife caused problems, as did having a lot of drunks on the Walt Disney World campus.

With the economy in freefall and Pleasure Island perceived as failing, Disney rebooted again. Almost incomprehensibly, they again prioritized theming. The story of the planned evolution of Pleasure Island into Hyperion Wharf is a tale for another day. Suffice to say that it was a placeholder for what came next. And that too was something with a theme. To Disney's credit, however, they finally learned from past mistakes.

The current theme of the region once known as Lake Buena Vista Shopping Village is...water. Technically, it is springs. Landscapers tasked with finding a more solidifying tie for the various areas of the former Downtown Disney sold their bosses on an obvious solution. The entire community was originally built with a natural centerpiece: a body of water that the first Imagineers carved out of land from Walt Disney's beloved Florida Project. By embracing the heritage of the land, they could augment and honor the legacy of their founder. Sometimes, the clearest solution is also the best one.

Disney Springs is the culmination of more than 40 years of attempts at perfecting the primary entertainment complex at Walt Disney World. Fittingly, The Walt Disney Company re-introduced their seminal entertainment industry in 2015, fifty years after Walt Disney himself looked at a map and said, "I don't care how long it takes. We have to buy that land." As always, he was right…and far ahead of his time with his strategy.

CHAPTER TEN
How Frozen Took Over the World

The announcement that Disneyland would no longer feature its beloved Twilight Zone Tower of Terror shocked die-hard supporters. The news that Disney would replace it with a themed Marvel attraction, Guardians of the Galaxy, added intrigue. This wasn't the first time that the company chose branding over legacy.

The turning point that led to the demise of Twilight Zone Tower of Terror didn't happen at Disneyland. It occurred at the second gate, at Walt Disney World. It was there that Disney killed a beloved attraction that was both engaging and amiable. They did so in favor of a much more popular brand, thereby returning to their roots as an extension of Walt Disney Animation Studios. This is the story that explains why Disney permanently closed Maelstrom in favor of Frozen Ever After.

Let's rewind to the early 1980s. In 1982, Epcot's World Showcase debuted to glowing reviews. Critics loved this permanent World's Fair. At the launch, nine countries participated, but park publicists promised that other pavilions would soon follow.

The first of them was Morocco, less than two years later. The second one and, in fact, the last addition to the World Showcase for reasons clear only to Disney execs, was Norway. In June of 1985, the *Orlando Sentinel* alerted the public to the existence of blueprints for what would become the 11th pavilion. The country of Norway believed in the project so much that they footed the bill for a key portion of construction.

Their parliament directed $10 million to the Norway Pavilion, expecting that a permanent presence at the Most Magical Place on Earth would boost American awareness of their country. They expected an influx of American tourists soon after the pavilion's debut, and they further projected that this stream of visitors would continue throughout the lifetime of the pavilion.

Whether those expectations were too ambitious is up for debate. What's inarguable is that Norway and Disney did everything they could to stack the deck. Norwegian businesses lined up for the opportunity to participate in this exciting venture. The Norway Pavilion opened in 1988, which also means that Disney hasn't opened a new pavilion in nearly 30 years.

The key attraction in the Norway Pavilion, Maelstrom, wasn't exciting, but it encapsulated several notable aspects of Norwegian culture. The most famous of them is the Viking mythos. Scandinavians celebrate the fables of Odin, Thor, and Ragnarok. The subject matter is fertile for ride development, and even though the World Showcase prioritizes accurate reflections of foreign cultures, the attractions display a bit of whimsy.

The first such attraction was El Rio del Tiempo, which laid the groundwork for the current iteration, Gran Fiesta Tour Starring The Three Caballeros. The original Mexico Pavilion water ride was the first of its kind at the World Showcase. A few years prior to the Mexican boat trip, Disney offered different World Showcase attractions with their Circle-Vision 360° movies of Canada, France, and China. The France film still exists today, while Disney worked with Canada and China to update their versions for the 21st century. All of them share the same weakness: they're glorified travelogues with a tourist sales pitch.

In an odd decision, the Norway Pavilion's new attraction would combine those themes. Maelstrom would offer a journey into the supernatural figures of Norse mythology. Theme park tourists would steel themselves for a trip down the path once trekked by actual Vikings. They'd face legendary creatures such as Dökkálfar and Ljósálfar, the Dark Elf and Light Elf of lore.

After the ride was over, guests would have a chance to watch a movie about the actual history of Norway, which involved fewer trolls than the preceding boat ride. The watery splashdown that signified the ending of Maelstrom seemed incongruous with the accompanying non-imagined documentary about the customs of Norway entitled *The Spirit of Norway*. It was largely a celebration of Norwegian winter sports options, which couldn't have less to do with Ragnarok unless there's a heretofore unpublished version of the Norse Bible. World Showcase fans enjoyed the process anyway.

The problem Disney faced with Maelstrom was simple. It wasn't the most ambitious ride when they made it in 1988. More than a quarter century later, the wear and tear on Maelstrom was unmistakable. Still, people loved it. The problem is that the sheer volume of people who loved it wasn't remarkable from a business perspective. The passionate support some folks demonstrated toward Maelstrom didn't translate to ride throughput.

Dropping the business jargon, Maelstrom boats were empty far too often, and few riders stuck around to watch the movie. The Norway Pavilion was no longer earning the tourist bonuses that they'd planned. Theirs was one of the only pavilions to offer a ride, but the benefit of that gradually eroded over time.

Fittingly, the central theme of Ragnarok is that the old regime collapses and a new entity rises to take its place. At the same time the demand for Maelstrom was dissipating, a pair of royal siblings was sneaking into movie theaters. Their presence seemed innocuous enough. After all, the first preview teaser for *Frozen* failed to include Anna and Elsa. Why would anyone expect them to become the most dominant force in the history of the Norway Pavilion?

That's exactly what happened. In November of 2013, *Frozen* dominated Thanksgiving week at movie theaters across America. That was only a small part of what it achieved during its theatrical run, however. The instant classic shocked box office analysts, myself included, by becoming the number one release of 2013 as well as the most popular animated movie of all-time, a fitting honor for Walt Disney Animation Studios to hold.

Frozen wasn't the film anyone would have expected to hold that title. It overcame modest projections to gross roughly $1.276 billion worldwide. At the time of its release, it was the fifth most popular movie ever released. And anyone who claims they saw that coming is lying.

The fallout was that *Frozen* became the dominant force for all things Disney in 2014. The company earned $4 billion from merchandising consumer products, and a significant chunk of that came from the princesses of fictional Arendelle. The Halloween of 2014 might as well have been a national Anna and Elsa cos-play event.

Disney quickly moved to boost the park presence of Anna and Elsa wherever possible, an understandable decision. I won't recount the debates about whether Frozen Fever negatively impacted Disney theme parks. Suffice to say that there was a lot of it, most of the gear sold, children experienced euphoria as they embraced Anna and Elsa during meet-and-greets, and strategists at the World Showcase noticed a huge surge in traffic.

What happened next was just as predictable as it was divisive. Park planners correctly deduced that a themed version of Arendelle was more befitting of the Norway Pavilion than the dated Maelstrom attraction. Also, without benefit of a Norway travelogue, the country was enjoying a spike in tourism. North Americans and even Europeans wanted to visit the lands that provided the backdrop for Arendelle. Yes, it was a made-up place, but the inspiration for it was the Norwegian landscape.

With Maelstrom no longer serving a purpose as a draw for Disney or for Norway, both parties understandably reached the same conclusion. A Frozen ride simply made more business sense. It was also an easy sell for Disney. From day one at Disneyland, themed lands were an indelible part of their history. They'd often base attractions on their iconic movie library. Why shouldn't they do the same for *Frozen*, one of their current breadwinners?

On September 12, 2014, Disney confirmed that Maelstrom would close permanently the following month. A few months later, the company confirmed that Frozen Ever After would replace it at the Norway Pavilion, creating a permanent tribute to fictional Arendelle in the otherwise authentic building. The outcry was immediate, and a #savemaelstrom hashtag campaign trended on social media for a time. It didn't matter, though. The money was in Anna and Elsa merchandising. The prior iteration of the boat ride never stood a chance.

Frozen Ever After debuted on June 21, 2016, and it immediately triggered a massive influx in traffic at the Norway Pavilion. Rather than build an entirely new attraction from scratch, Imagineers retrofitted much of the Maelstrom boat technology into the updated version. They saw an opportunity for a quick turnaround from a dated attraction to a themed one that would create new excitement in an unpopular part of their park. From a business perspective, it was the only choice.

Notably, this decision was one presciently endorsed by Walt Disney himself. He famously stated the following: "Disneyland will never be completed. It will continue to grow as long as there is imagination left in the world." Obviously, the same logic applies to Walt Disney World and Epcot in particular. Disney himself built attractions for the 1964 New York World's Fair, the most famous of which was a themed boat ride, demonstrating that he'd happily have signed off on Frozen Ever After.

The problem Disney fans face is the constant change to perform such actions. We all have a memory of our first time at a certain Disney attraction. When the ride closes, we feel as if we're losing one of our favorite recollections. For anyone who visited the Norway Pavilion prior to 2015, Maelstrom embodies such remembrances. Its loss is deeply personal.

The same is true of Twilight Zone Tower of Terror, one of the greatest attractions in Disney theme park history. Guests at Disneyland will no longer have the same memories because the company replaced it with a Disney-owned intellectual property. The new theme is *Guardians of the Galaxy* film, which was notably the surprise hit of 2014 on the heels of *Frozen* being the surprise hit of 2014. And Disney put it into that repurposed area rather than develop in a new space at the park.

In other words, history is already repeating itself. *Frozen* built a playbook, and the company has enough confidence in it to employ the same tactic with *Guardians of the Galaxy*. Whether it works as well remains to be seen, but the turning point is unmistakable. The company is circling back to themed attractions of Disney licenses at the cost of beloved stand-alone rides like Maelstrom and Twilight Zone Tower of Terror. The economics of the choice are perfectly reasonable. It's the human element where Disney fans are feeling the sting of loss.

CHAPTER ELEVEN
The Fifth Disney Park

From a young age, people learn to think of things in factors of five. For whatever reason, teaching is easier through the "5, 10, 15, 20..." method. Perhaps it's this aspect of our nature that causes so many people to think of Walt Disney World in such terms. Why would anyone stop at four parks when five is so much easier to wrap your head around?

The problem with this line of thinking is that it puts your entire perspective of Walt Disney World in a state of limbo. After all, Disney hasn't added a new park in Orlando since 1998. That's 19 years and counting since something new debuted at the world's most popular amusement park. To put this into perspective, consider that Epcot, Hollywood Studios, and Animal Kingdom all opened during a 16-year span from 1982 to 1998.

For all the speculation about the fifth park, Disney keeps unveiling new ways to expand the parks that actually do exist. Animal Kingdom famously received a Pandora: The World of Avatar expansion in 2017. During the summer of 2015, Disney finally confirmed the long-standing rumor that Star Wars and Toy Story expansions are in the works for Hollywood Studios.

That's a lot of enhancements for the two less-popular parks at Walt Disney World. It's not the news that people have anticipated since 1998, though. Everyone knows that Walt Disney covertly acquired over 43 square miles of land during the 1960s. Only a third of that space has been developed thus far. There's still plenty of room for growth, and Disney owns so many intellectual properties that they could easily fill several more parks if they choose to spend the resources doing so.

With the Harry Potter license continuing to pay dividends at nearby Universal Studios, Disney does appear to recognize that they have to up their game in an increasingly competitive marketplace. Until they announce a new park, however, all that we can do is speculate on what might have been if any of the rumored fifth parks for Walt Disney World had been built. Here's everything you need to know about a few (apparently) mythical Disney theme parks in Orlando.

There was a rapid expansion of Walt Disney World from Magic Kingdom in 1971 to a four-park mega-tourist destination by 1998. Given the epic growth and the amount of land The Walt Disney Company owned in the greater Orlando area, it's understandable why people immediately started speculating about what would come next. At the turn of the millennium, no one could have anticipated that 15 years later, Walt Disney World would still host only four parks. They'd added one every half a dozen years on average since 1982. Why would the company move away from such behavior?

By 2001, people already had a firm belief about what would come next. During the early days of the internet, a few people enjoyed strong relationships with Disney cast members, including Imagineers. Before the concept of the internet scoop was a part of the zeitgeist, some of these people would post details about impending Disney theme park projects. Their success ratio was intermittent at best, but that's not a reflection of whether their information was correct. The Walt Disney Company often enters the discussion and even the planning stage for an idea before dismissing it as impractical. This is true even today, which is why one of the parks you'll read about is no longer a stand-alone entity.

The point is that in the days before social media existed, internet readers took note of anyone who had viable connections to Disney employees. People who successfully projected future theme park plans garnered respect and credibility.

In August of 1998, Disney's Animal Kingdom was barely three months old. The *Los Angeles Times*, a newspaper that had chronicled Disney theme parks since day one and even a couple of years prior to that, investigated the company's future. They acquired an important quote from someone who would know about Disney's impending theme park plans.

During most of the 1990s, Judson Green, then-president of Walt Disney Attractions, held a title equivalent to being leader of Walt Disney Parks and Resorts today. He determined the overall direction of Disney theme parks. When pressed for ideas about a fifth gate at Walt Disney World, Green offered this thought-provoking response: "It's really premature to talk about a fifth gate at Orlando. The fact of the matter is we are only beginning to think what that theme might be."

In other words, as early as 1998, Disney already had a series of plans for its follow-up to Animal Kingdom. Almost 16 years later, there's still no fifth park. One of the reasons for this is that then-CEO Michael Eisner famously experienced several setbacks in dealing with key Disney personnel at the upper levels of the company. Green himself left The Walt Disney Company in 2000. A 19-year veteran, his exit created additional headlines about the problematic direction of Disney during that era.

While Green did achieve his primary goal, setting the table for the expansion of Disney theme parks into China, even that park, Hong Kong Disneyland, wouldn't open for another five years after his departure. All plans Green had for the fifth gate fell by the wayside after he left. Coincidentally or not, they leaked during the year that followed his exit.

LEGOLAND Florida broke ground in 2010, and it debuted in 2011. A full decade before the park opened its gates for the first time, however, Disney blogger Jim Hill reported that it would take on a different form. At the time of the rumor in 2001, there were already three LEGOLAND parks in existence, with a fourth opening scheduled in Germany the following year.

Even though LEGO Global Family Attractions was dealing with a relatively new park in San Diego and an upcoming project in Günzburg, they jumped at the opportunity to discuss a joint venture with executives from The Walt Disney Company.

Hill chronicled high-level meetings that transpired in Burbank, California, between two of the most powerful family-friendly companies in the world. LEGO, one of the strongest companies in Denmark, structured an ambitious plan for the growth of their company. One of the most daring involved building a park with the LEGO name as part of Walt Disney World.

Your first question is undoubtedly why Disney would like the idea. That's easily explained. In 2001, the company was in the middle of tumult that eventually led to the resignation of Roy E. Disney, nephew of Walt Disney and son of Roy O. Disney. Roy E. represented the connection to the glory days of the park. By the turn of the millennium, he was feuding with CEO Michael Eisner, who was notoriously frugal. Roy E. Disney would never consider a Disney theme park that didn't have the stamp of his family's legacy on it.

Eisner was less concerned with the history of the Disney family. His focus was on the share price of the stock and the net revenue of the company that impacted it. He loved the idea of a theme park that required little heavy lifting from Disney employees to bring to life. The cost of a LEGOLAND Park at Walt Disney World made too much financial sense not to explore, at least to someone who wasn't attached to the Disney legacy.

The financial struggles of Animal Kingdom were the driving force behind Eisner's thinking. The park failed to meet its pre-opening projections for attendance and revenue. And it missed by *a lot*. After spending over $800 million to build the world's coolest joint theme park/zoo, Disney's Animal Kingdom suffered a great deal of negative publicity during its early days.

Also, people weren't quite sure what to think of the historically unprecedented merger of concepts. A lot of potential customers waited to hear what the reviews were for Animal Kingdom. Since those evaluations were middling, park traffic disappointed Disney executives. Disney would need to invest another $200 or $300 million to lure enough visitors to bring Animal Kingdom up to its initial projections. That would make the park an investment of over a billion dollars simply to match original expectations.

This situation explains why the plans that Disney was considering for a fifth gate as early as 1998 fell by the wayside. At the time, they simply didn't have the money to spend on a new project while the most recent one struggled. LEGO presented them with a remarkable plan for park that would be cheap to build. LEGOLAND California had cost only $130 million, and Disney would be splitting the costs with a different company. Even at full price, however, the amount they'd need to invest to add a LEGOLAND gate to Walt Disney World would cost less than the investment they'd have to make to bring Animal Kingdom up to snuff.

On paper, this plan looked great. LEGO and Disney are two of the most popular toy manufacturers in the world, LEGOLAND had proven itself popular in Europe, its California reception was better than expected, and they could build a new park for much cheaper through a joint venture. There were only a few problems. The most obvious one would be determining who was in charge. Eisner was a notorious control freak, and he wasn't about to cede power at one of his company's crown jewels to the leaders of a different company.

There was also the issue of cost. Even at a cheaper total than building a new park, Disney execs weren't convinced that the new addition would add enough traffic to justify the public relations hit they'd take for allowing a non-Disney (but officially licensed) park on Disney grounds. LEGOLAND California claimed less than a quarter of the traffic of Animal Kingdom, the least successful of Disney's four parks at the time. While two million annual guests is wonderful for a non-Disney theme park, that type of bump wasn't enough to move the needle on Disney's dial, even in 2001.

Perhaps the determining factor, however, was the decidedly low-tech design of LEGOLAND theme parks. Even Disney's Imagineers would struggle to make such cheap set pieces seems worthy of the Disney park brand. They ultimately decided that working to improve what they had was better than introducing another potentially sub-standard park on the heels of Animal Kingdom. In the process, they broke the hearts of many unknowing Duplo fans.

For their part, LEGO remained adamant that they should continue to build their brand via theme parks. They've admirably done exactly this, expanding their park holdings. They even found a spot to build their own park in Orlando at the site formerly used by Cypress Gardens. Also, LEGO as a brand has become so popular that LEGO thievery became a huge industry for some enterprising criminals. And *The LEGO Movie* earned almost $470 million worldwide, proving that LEGOs are a part of the mainstream now.

The good news for Disney was that they managed to redeem Animal Kingdom to the point that it has supplanted Hollywood Studios. Since its shaky start in 1998, attendance has increased. Re-investing in Animal Kingdom made perfect sense at the turn of the millennium, and it continues to do so, which is why Pandora: The World of AVATAR had its debut in 2017.

The two companies continue to enjoy a mutually beneficial business relationship. Disney LEGOs are among the most popular sets each year. Disney even trusts LEGO with its signature brand, the Princess Collection. A theme park union of the two companies would have been amazing to watch, though.

For several years, there were three popular rumors about the fifth gate. The most prevalent of them is the one that makes the most sense. Since the introduction of *Snow White and the Seven Dwarfs* into pop culture in 1937, no one has created more memorable villains than The Walt Disney Company. I mean that in terms of quality and quantity. Beginning with the Evil Queen Grimhilde, we could spend the rest of this chapter describing memorable evil-doers and why they've stood the test of time in the memories of Disney fans. We're obviously not going to do that, but let's hit the high points.

No list would be complete without mentioning Maleficent, first introduced in 1959's *Sleeping Beauty*, and then brought into the realm of live action in 2014 when Angelina Jolie portrayed the titular lead. Captain Hook shows up in every Peter Pan movie, and he's also quite popular on ABC's *Once Upon a Time*. Gaston from *Beauty and the Beast* is more than just a hilariously egomaniacal foil for Belle. He's also a viral video sensation thanks to the great cast members who portray him at the parks. Cruella De Vil doesn't just have devil in her name; she's also the source of inspiration for one of the greatest Simpsons songs ever, "See My Vest." Then, there's Scar, who single-handedly sets back familial relationships among the kings of the jungle through his actions in *The Lion King*.

Finally, if we bring the concept up to date through a recent Disney purchase, think about Star Wars from the perspective of cult of personality. Luke Skywalker was never as popular among fans as Han Solo, the ne'er-do-well imprisoned by Jabba the Hutt due to poor relationship choices. His lack of team spirit is right there in the name, yet he eventually joins forces with the good guys to stop the Evil Empire. And no matter how popular Solo is, no matter how much debate he provokes about who shot first, Han Solo is nowhere near as established in the pop culture zeitgeist as Darth Vader.

As much as anyone else here, Vader exemplifies how people perceive villains. Walter White never becomes a cultural icon unless he develops the Heisenberg persona. Nobody cares about Dr. Jekyll until he devolves into Mr. Hyde. John McClane is an angry man in danger of becoming a weekend dad before Hans Gruber invades Nakotomi Plaza. Superman isn't interesting as a virtually impervious hero without Lex Luthor constantly trying to find ways to make him vulnerable. And Batman is merely an uptight billionaire with anger management issues without the Joker cracking jokes while wreaking mayhem on the suckers still inexplicably living in Gotham.

In every realm of storytelling, villains differentiate and elevate heroes. Since Disney has the world's greatest catalog of evil-doers, their rogues gallery represents a competitive advantage in the theme park industry. By monetizing the bad guys, they could theoretically double sales. Think about it from a different perspective. A product that appeals exclusively to women excludes 49 percent of the population while one catering to men is even worse; it excludes 51 percent of humanity.

The same rationale applies to merchandising sales, Disney's bread and butter. Selling Lion King toys without releasing a Scar brand is a mistake. They miss out on a key component of their revenue stream. We know this from Star Wars sales. A recent study of the revenue generated by each character determined that Darth Vader leads the pack. Amusingly, a member of the Dark Side also finishes in second. Boba Fett appeals to more people than R2-D2 when they vote with their wallets. Disney figuratively leaves money on the table every time they ignore potential villain revenue.

Disney has recognized this for a long time. Even before they owned the Star Wars license, they had their own data that suggested they should merchandise bad guys more often. The problem is that performing this action requires a fundamental change in behavior after more than half a century of sameness. They would have to do something bold to display their villains in the most buzz-worthy manner possible. The Walt Disney Company realized exactly what they needed.

Rather than start marketing a wide variety of merchandise for villains seemingly out of nowhere, Disney came up with a great plan. They would build an entire park full of evil doers, and then they would capitalize on the word-of-mouth to launch an entirely new line of products. It was exactly the kind of visionary thinking predicated upon potential revenue increases that make Disney a Fortune 100 company in the first place.

Once they started evaluating the issue, however, a few holes popped up. Since literally the first day at Disneyland, children had struggled with the scarier elements of some rides. Initially, it was Snow White's Scary Adventures that frightened children. They didn't enjoy the too life-like witch, which is to say that she absolutely terrified many kids. To a certain extent, that's the idea of putting the bad guys on display for everyone. It's supposed to add an extra burst of adrenaline and a bit of entropy to the experience.

A park featuring Disney's rogue's gallery could offer more thrill rides and more excitement than, say, the basic attractions at Magic Kingdom. They could also use the world's most popular theme park as a basis for an antithetical location. If a Disney princess would have her own castle as the central point of Magic Kingdom (Cinderella Castle) and Disneyland (Sleeping Beauty Castle), logic dictates that one of their enemies would inhabit a castle at Dark Kingdom.

The plan was for Maleficent to lord over the villains as the hostess of Dark Castle, a sort of mirror universe concept. The park would offer twisted versions of established Disney attractions. If Ariel has an Under the Sea ride, it's only fitting that Ursula offer a nefarious counterpart. Dark Kingdom plans included mountain peaks matching the famous ones at Magic Kingdom, albeit with a dark spin. Think of them as evil mountain ranges with Disney villain themes.

In the end, Disney couldn't justify a new park with such a bold premise. They would have to risk too much for potentially little gain if the idea didn't prove mainstream enough. Eventually, they rebooted the plan a bit as Shadowlands, an extension within Magic Kingdom that would include some of the same premises. It didn't seem like a great fit with the kid-friendly themes of the rest of the park, though.

The idea of a twisted take on Disney's existing theme parks seemed wonderful on paper. They could boost merchandising sales on properties that hadn't been mined properly in the past. The social media era and its revelation of the power of niche products emphatically proved this as a good idea. But Disney chose to populate their theme parks with bad guys in a more general sense rather than build an entire park. Mickey's Not-So-Scary Halloween Party included a Hocus Pocus Villain Spectacular, a returning favorite from past years. Also, during May of 2014, the Memorial Day 24-hour celebration Rock Your Disney Side included a villainous pre-parade. Disney even added a Club Villain experience at Hollywood Studios. At a hefty price of $99 per person, people could enjoy dinner and dancing with evildoers such as Cruella De Vil, Maleficent, the Queen of Hearts, and Dr. Facilier. Clearly, this enchantment with the dark side isn't going away anytime soon, and I hope that Disney eventually brings Dark Kingdom to life.

Another rumored park, the Night Kingdom, was to be an after-hours facility that would admit only 2,000 guests daily (well, nightly), similar to SeaWorld's Discovery Cove.

Jim Hill wrote about the Night Kingdom in 2008:

> And, yes, the Mouse *is* actually going to build this $520 million project. Current plans call for this niche park to officially throw open its doors in October of 2011, just in time for the start of Walt Disney World's 40th anniversary celebration.

I'm not looking to call Hill out for his faulty prediction. I'm simply pointing out how likely this premise seemed several years ago prior to falling apart completely. You can already tell one negative about it. Magic Kingdom enjoyed 19,332,000 visits in 2014, an average of approximately 53,000 guests each day. A park allowing admittance to only 2,000 would mean that over 96 percent of all visiting guests at Walt Disney World would leave disappointed.

There's also a financial issue. Night Kingdom would cost $520 million to build. That's roughly $260,000 per daily guest. Even over the course of a decade, Disney would need a profit margin of $72 per guest simply to break even, and that's before we include daily operating expenses. The numbers on this one never made any sense, which is why it was all the more surprising that Hill seemed so confident that the plan was moving forward.

In 2008, SeaWorld was garnering a lot of attention for Discovery Cove, but they have wholly different site traffic considerations than Disney does at Walt Disney World, far and away the most popular theme park in the world. It was obvious to anyone considering the project that the 2,000 daily limit would never work, for several reasons. Why, then, was it even considered?

Night Kingdom would operate from 4pm to midnight, providing guests with an unprecedented park experience. Disney intended to hire 4,000 cast members—yes, two for every one site visitor—to create a velvet rope type of atmosphere for their most exclusive park. Unprecedented features such as zip lines over crocodile dens, hand feeding of live animals, and even a penguin playdate were planned.

Of course, all of these amazing features, ones Walt Disney couldn't have imagined when he was building Disneyland, came with a hefty price tag that would give ordinary people extreme sticker shock. In a way, it was the forerunner to Disney's current operating practices at parks. By catering to guests with expensive options, they could entice the "one percent" to spend boatloads of money at a place where they wouldn't have to interact with the standard Disney rabble. They would even enjoy a special Broadway type of show that would be the finest of any Disney theme park, rewarding the higher paying clientele.

Night Kingdom was also a forerunner for an idea that did become real. One of the strongest selling points for the swanky park would be a nighttime safari similar to the Animal Kingdom's Kilimanjaro Safaris experience. The difference is that all the nocturnal animals that sleep during the day would be on display after dark, something a regular Animal Kingdom guest could never experience at the time.

Of course, you probably know that Disney later added a nighttime safari for guests staying at Animal Kingdom Lodge. Entitled Sunset Kilimanjaro Safaris, it become available to everyone as the park moved away from the half-day experience with the introduction of Pandora: The World of Avatar. They also added Rivers of Light, a nightly show with fitting nature themes. They even added some of the concepts such as standing above a crocodile den into Animal Kingdom. The difference is that there's a rope bridge instead of a more dangerous (and more expensive) zip line.

In other words, the original concepts for Night Kingdom were largely grafted into Animal Kingdom improvements instead. Once again, Disney chose to improve the quality of their fourth gate rather than expand to a fifth one.

The most famous rumored fifth gate ties together several of the themes above. It's an expansion that would be cheap to make, it would be a boutique park, and it would highlight an already discussed Disney villain. That park is obviously Star Wars Land, which you've assuredly heard is going to become a part of a massive update to Hollywood Studios.

During its original planning stage, however, Star Wars Land would stand on its own. The Disney acquisition of Star Wars from George Lucas in 2012 was an exciting development on several fronts. It meant that they could continue the movie franchise, of course, but it also gave Disney the keys to one of the best-selling merchandise machines that they didn't already own. Star Wars instantly flipped from being the competition to standing as one of Disney's linchpin intellectual properties.

The company spent months discussing the best ways to monetize the license, something they're still doing today, and a theme park was a logical conclusion. After all, Star Tours has remained a consistent draw since its Disneyland debut in 1987. When they reinvented it with multiple branches in 2010, Star Tours: The Adventure Continues proved that the icy reception of the prequel movies had done little to besmirch the overall popularity of Star Wars as a franchise. If Jar Jar Binks can't destroy you, you're invincible.

Like many Star Wars products before it, Star Wars Land would mercilessly punish its devoted fanbase by charging them to live out their dream. The idea was of a Jedi Training School, not to be confused with the children's show they run at Hollywood Studios. For a measly $200, a person could enter the park and feel like a new recruit learning the ways of The Force.

Like Night Kingdom, Star Wars Land would remain open fewer hours than standard Walt Disney World parks. So, a price more than double the cost of Magic Kingdom would make even Comic Book Guy choke on his 100-taco meal. Even to single techies with tons of disposal income, the charge is outrageous. You have to *really* want to be a Jedi Master to pay that kind of price.

How did Disney intend to justify the cost? Have you ever heard of fantasy baseball camps? They're the ones where people get to spend the weekend training with their favorite players. Star Wars Land as initially planned would work a lot like that. A customized menu of options would allow for heavy personalization. With technology advancing at a rapid rate, Disney intended to give people the type of Jedi training they desire to make them feel like they really accomplished something by the time they left the park. In that regard, Star Wars Land would have represented a unique opportunity for some of the most loyal fans in the world to live out their dreams, at a dazzling cost.

As you know, Disney eventually decided to go a different way with the concept. The Star Wars Land they've announced is a much more conventional theme park experience featuring new rides and character interactions, albeit ones people who visit during Star Wars Weekends already know. Don't get me wrong. The presence of an actual Star Wars area at Hollywood Studios guarantees a crowd increase so significant that I can easily envision it overtaking Epcot to become the second most popular theme park at Walt Disney World. Keep in mind that it is currently last, 90,000 guests behind Animal Kingdom and over 1.1 million guests behind Epcot.

So, I'm projecting a massive spike in park traffic the instant Star Wars Land opens. It's just unfortunate that something with so much promise got reduced rather than expanded. When Disney purchased the Star Wars license, nobody was hoping that they'd do just enough. People want Disney to shoot the moon on this particular franchise, and I'm dubious that the current iteration of the concept will satisfy guests.

Here's the amazing thought about Disney's stagnation during the timeframe from 1998 until today. Islands of Adventure at Universal Studios is newer than any theme park at Walt Disney World. And it's already experienced a major upgrade with the Wizarding World of Harry Potter. This lack of progress explains why Disney is so focused on improving Hollywood Studios with Star Wars Land and Toy Story Land expansions. Their current setup is remarkably dated relative to the historical progress of Disney theme parks as well as the development of others across the world. Consider that Animal Kingdom is only six years newer than Disneyland Paris, a park that is notoriously run down by Disney standards.

Disney's failure to build a fifth gate continues to frustrate customers. Many observers expected them to finally announce a new park in 2015, but the company chose the boosting of a current Walt Disney World property instead. Will there be any new developments in coming years? If there are, nobody is saying anything yet. Well, maybe Jim Hill is, but since he's responsible for almost all the rumors above, we'll be a bit dubious about the next thing he suggests. Sorry, Jim.

In fact, Disney seems perfectly willing to improve what they have rather than expand beyond their current borders. Given their record-setting attendance in recent years, it's hard to argue against this strategy, frustrating though it might be. Still, with so many popular intellectual properties and so much Orlando property, it's undeniable that Disney will eventually open their fifth gate. The question is what form it will take and whether any of the ideas above are explored once more.

CHAPTER TWELVE
Disney Myths and Legends

The propensity to inspire legends is one of the strong suits of Walt Disney theme parks. Some of them are whoppers, too. Doomed spirits haunt some of the happiest places on Earth. Dolls come to life at night, switching places before the next morning's guests arrive. Walt's head resides somewhere on the premises. Or maybe he's not dead at all. Maybe he unearthed the secret of longevity, and that allowed him to take up permanent residence behind the scenes, controlling all phases of the Disney empire to this day.

Clearly, there are some whoppers out there, and not everyone thinks to check Snopes to ascertain the truth about Disney theme parks. What follows are a series of rumors that seem so far-fetched that no one could possibly believe them, yet some people do. Perhaps you've even fallen victim to a couple of these hoaxes yourself. If so, you'll want to read on to discover these completely untrue "facts" about Disney theme parks. Warning: there's a lot of debunking ahead. On the plus side, if you've got friends who spend an inordinate amount of time passing along this nonsense as a way to talk smack about Disney, you're about to turn the tables on them in fairly epic fashion.

Since its inception, eBaum's World has had a reputation for being a bit shaky with the truth. After a media conglomerate purchased the site and fired all the original staff members, including the site founder, this destination web hub became notorious for, let's say, borrowing content from other locations. The danger with that is that you occasionally cite someone else's nonsense as fact.

That's the only way to explain their article that claims a Disney cast member performs a nefarious action while guests are flying on Aladdin's Magic Carpet Ride. According to their information, someone controls the camel spitting. I guess the premise is that Disney figures it's a great idea for one of their employees to antagonize guests with a splash of water, especially if it's well-aimed enough to hit someone in the eye.

Think about the above for a moment. Is The Walt Disney Company in the business of antagonizing park guests? Sure, a flick of moisture isn't the most irritating thing that'll transpire during a theme park visit, but it's annoying nonetheless. Disney employs the water splash as a fun way to add an element of surprise to the festivities. If they added any element of meanness to the act, guests would complain. When that happened, Disney would stop doing it immediately. They built their entire empire around the concept of positive reinforcement.

To a larger point, Aladdin's Magic Carpet Ride is an automated attraction. Everything works on a timing system, which means that all splashing occurs in a way that you can actually anticipate if you study it enough. Of course, you don't even need to put that much effort into it. Watch the camel's head. That way, you'll know if and when any water is heading your way. There's no evil cast member, folks.

Disney's family-friendly reputation causes some folks to take every concept to its logical extreme. They believe that Walt loathed the idea of alcohol in his parks. There's actually a kernel of truth in this. Disney disliked the smell of beer, especially when it dried on hot concrete. Given the arid conditions at Disneyland, it made sense to disallow alcoholic beverages inside the park.

There were exceptions almost from the beginning, though. In 1967, a dozen years after the park opened, Club 33 debuted in New Orleans Square. Disney designed this lavish club for the patricians rather than the hoi polloi, and catering to such an elite crowd demanded the sale of wine and champagne. So, alcohol has been available on the Disneyland premises for almost half a century now.

The situation at Walt Disney World is a bit murkier. The subdivision of theme parks caused a divide in the standard rules across the sites. Epcot features several pavilions that include restaurants and bars. They couldn't accurately reflect the culture of Germany without beer, and I say that as someone whose given name is technically Mombauer. My people love their potent potables, and Disney respected that with their park dedicated to world culture.

Magic Kingdom played by very different rules until recently. It's true that the same original rules for Disneyland applied to Magic Kingdom for decades. The situation didn't change until the introduction of Be Our Guest Restaurant in 2012. In terms of publicly accessible Magic Kingdom establishments, it was the first to offer beer and wine to guests. Some guests were outraged at the time it opened, but what was effectively a trial run for public alcohol consumption at Magic Kingdom succeeded. In fact, the Jungle Navigation Co. Skipper Canteen, which opened in 2015, sells alcohol as well. Just please don't spill any beer on the concrete. It would make Walt roll over in his grave.

The legacy of Walt Disney is so epic in scope that rumor-mongers have devoted countless hours coming up with seemingly plausible ways that his spirit might live on at his parks. And those are the well-intended people starting those rumors. Haters craft even taller tales regarding his corpse and whether it may be a Corpse-sickle...Corpsicle?

Anyway, both parties wound up creating equally alarming versions of the truth. Some people believe with every fiber of their being that the frozen heads of Walt Disney and Ted Williams are resting in a vault somewhere, waiting to become real-life *Futurama* characters. Biography.com eventually devoted an entire article to debunking this belief. They note that the cause of confusion is a second party account of Disney's philosophy on the subject as relayed by a devout disciple of the corpsicle industry. And even *that guy* plainly stated that Walt's family cremated his body. So, if a friend ever tries to state this as fact, please do me a favor. Walk over to your refrigerator, open your freezer, grab a Popsicle, walk over to your friend, and smack them with it. They deserve it.

Disney's most ardent supporters don't do his cause any better, though. No, they don't think that the frozen remains of Walt Disney reside under Pirates of the Caribbean. Their belief is even more outlandish. They subscribe to the notion that Walt Disney loved his theme park so much that he couldn't leave it. After his death, he found a way to stay at the Happiest Place on Earth rather than shuffle off this mortal coil and ascend to the afterlife.

Yes, this sounds crazy. Yes, it *is* crazy. It's a real thing that has enough supporters to maintain an oddly large amount of traction. In fact, someone brings it back to the public eye again every few years. A recent example occurred in the winter of 2015. The Mirror and the

Huffington Post websites both ran stories suggesting that cameras accidentally caught a ghost on camera. At Disneyland. And it was Walt.

The whole thing feels like a campfire tale run amok. Someone with a vivid imagination doesn't know when to quit. But these posts are true. Some writer stumbled on a grainy 2009 YouTube video. It purportedly reveals security camera footage of a ghost trekking through, fittingly, the exterior of Haunted Mansion on its way to Rivers of America. The fact that it's obviously a reflected light doesn't seem to dissuade anyone from their Fox Mulder-esque belief. This 63-second trifle of viral video has somehow accumulated over 13 million views.

What you can conclude from the above is that people really want Walt Disney to haunt his theme park. Why they don't want him to rest in peace or enjoy his heavenly reward is beyond me. The strange part is that this isn't even the most desperate attempt to place Walt at Disneyland right now.

One of the least likely myths in the entire world, and almost certainly the quirkiest at Disney, is that Imagineers are even greater than you thought they were. During their research, they unearthed the technology to make Walt Disney immortal. I realize you're waiting on a punchline right now, but what could I say that's funnier than that?

The philosophy surrounding Disney's continued and not-at-all-impossible existence is that he wanted to get out of the spotlight. Rather than place himself at the forefront of all transactions involving The Walt Disney Company, the founder and namesake of the corporation instead resides in an understated apartment in Magic Kingdom at Walt Disney World. From this home hidden in plain sight, he orchestrates all the business decisions that allowed his people to buy Pixar,

Star Wars, and ESPN, among other financial master strokes. I strongly suspect that current CEO Bob Iger *hates* this myth, because it makes him the Milli Vanilli of the Fortune 100.

The support for the belief involves lighting. Some people claim that the candle in the window of Walt Disney's apartment on Main Street at Disneyland occasionally turns itself off and on. Yes, that candle is always supposed to stay lit. Its flickering or turning off means a power disruption, not a supernatural event.

Despite the obvious explanation, there's a wives' tale about why Disney chose to leave this light on. A maid allegedly tried to turn off the light in Walt's Disneyland apartment after his death. After exiting the residence, she noted it was still on. Presuming she had forgotten to turn it off, she went back in specifically to do so. To her shock, the instant she exited the building again, the lamp activated once more.

The terrified cast member relayed the incident to others, and that's supposedly why Disney employees chose to leave a light on for Walt. Reading between the lines, someone else on the custodial staff played a fantastic prank on a co-worker, she took it seriously, and now Disney has to waste energy powering a lamp. This must be Ashton Kutcher's favorite Disney myth.

To a larger point, if such a magic elixir existed, Disney's Nine Old Men really could have used it. Why did Disney hog it all for himself and let the Imagineers who discovered its magical properties wither away via the natural aging process? He could have saved a fortune in training new Imagineers by keeping the old guard, the originals, instead. Despite the gaping logic flaws with the Immortal Walt rumors, it persists in many circles. To wit, if you google "Walt Disney still alive," there are 73,000 results. People really want to believe that whether by head-in-a-jar or scientific witchcraft,

Walt bested Father Time. If he does come back to life and all these rumors prove accurate, feel free to lord this article over me.

According to urban legend, the godfather of Disney theme parks isn't the only spirit who walks the streets at night. Indeed, Walt Disney World and Disneyland are two of the greatest inspirations for ghost stories. It's presumably because so many people try to spread the ashes of their loved ones at these two parks. With that many parts of the deceased floating in the air, someone's going to do some haunting.

The most popular recent story involving spirits at Walt Disney World allegedly occurred at Twilight Zone Tower of Terror. A Disney cast member in full ride regalia as a bellhop tested one of the rides prior to opening. You probably know that there are multiple loading platforms for this ride, and they're cleverly named Alpha, Beta, Charlie, and Delta. Echo and Foxtrot are the shafts directly below where the ride gets exciting.

The bellhop received the assignment of checking the Delta platform. That's apparently the unlucky one, because story goes that this employee dropped dead of a heart attack while standing on the platform. Rather than leave the earthly realm behind, this dedicated Disney worker chose to remain on Delta forever. He now haunts the facility, providing a spookier ride experience for it than on the other three platforms.

Nobody loves this myth more than Disney cast members. It adds an element of excitement on dreary days when the lines are long and the customers are irritable. Plus, it's much easier to get into character if you actually believe that people could die on the attraction. The ghost will even knock off the hat of any cast member who shirks their duties, apparently.

The social media era has created an unfortunate by-product of fake rumors, videos, and images. The idea is to build a story credible enough that it doesn't fail the laugh test. If you can scare people enough to give them a chill or raise the hair on their arms, you'll get a lot of attention.

The Haunted Mansion is a perfect candidate for this sort of behavior. The turbulent history of the attraction and its strong association to several Disney theme parks guarantee that fans of the macabre will always feel drawn to it. They tether their own mythologies to existing ones, pushing the boundaries toward the scarier side of an often humorous ride.

The most famous recent example is a ghost tale surrounding a pair of boys, one of whom laughs continuously while the other one cries. There isn't a lot of backstory as to how or why the kids wound up trapped for eternity at a Disney theme park. The fact that one cries all the time is odd, because I've just described heaven for a lot of children. Anyway, a commenter on a Disney forum relayed his encounter with one of the boys and described his confusion when he studied a picture he'd taken while riding a Doombuggy:

> I took my 100-foot night shot accessory to WDW with the specific purpose of taking ride photos of the Haunted Mansion. Well, after documenting the ride, I put the camera away for the rest of the day. I went back to our place that evening and began to download the photos to my laptop. After doing so, I scrolled through them to make sure I had all I needed, and lo and behold, on one of the first shots of the attraction, I saw something that definitely WAS NOT there when I was on the ride itself. This photo was taken in the first hallway of the attraction with the eye-following portraits.
>
> It appears as though a child is peeking his head out of the Doom Buggy and looking directly at me. Not

only was he not there when I took the pic, there wasn't a child of this age within 20 people in front of me in line, and as you can see, he's only a few Doom Buggies in front of me. Not only that, what's he doing looking at me? There is NO flash, and NO visible light coming from me. It's all infrared, and invisible to the naked eye.

Despite this convincing tale, there's no proof that this was anything more than a bored child he didn't notice a couple of Doombuggies ahead of him. Because ghosts aren't real, right? *Right?*

His name was George, and he was a die-hard employee of The Walt Disney Company. A welder by trade, George showed up for work every day, did his job, and made his co-workers smile. His dedication to his craft almost supersedes the fact that he didn't exist. Tragically, George's job requirements included climbing a set piece from the early days of the attraction. It was called the Burning City, and it stood high in the air. As George attempted to climb up to service the contraption, he tripped and fell to his death.

George was an amalgam of two different Disney construction workers who experienced tragedies during the early days of the park, neither of whom were stationed at the work site for Pirates of the Caribbean at Walt Disney World. The honesty between the lies is irrelevant to the myth, though. Twisters of truth remember George the welder as a single man who later became a spirit called the Ghost Pirate of the Caribbean.

Disney cast members have offered their respects to George for decades now. They understand that if they fail to greet him over the PA system in the morning and at night, they'll run the risk of incurring his wrath. George is equal parts friendly ghost and hostile spirit. People who befriend him will enjoy an easy work day.

Those who fail to acknowledge him will suffer ride breakdowns, power outages, and even the occasional ride injury.

Basically, George receives the credit for every incident that occurs on Pirates of the Caribbean. Many current and former cast members relay personal anecdotes about weird things that happened on the ride that they believe are otherworldly in nature. In reality, this is a rickety old attraction that takes place in water in the dark. Strange things occur, and none of them involves a ghost whose name people don't even remember correctly. On the plus side, it gives cast members something to look forward to each day as well as something to blame when things go wrong.

Should you have the courage to google suicidemouse.avi, well, you'll be in for a disappointment. It's an elaborate hoax involving the mouse that started it all. Some enterprising viral video creator recognized that spinning the beloved icon into a dystopian victim would be brilliant years before Banksy borrowed the idea.

Mickey Mouse Goes to Hell was an early contender in the found footage genre. The premise was that an executive (in some versions, the archivist is noted film critic Leonard Maltin) discovered hidden footage on a previously released animation clip. The video only displayed the first three minutes of a nine minute production. Curious, he tried to watch the rest and basically wanted to gouge his eyes out by minute seven.

The clip's audio alters around this point, evolving into a gurgling noise that sounds like someone drowning. After that, a screaming sound emerges. Meanwhile, an unholy version of Mickey Mouse appears onscreen. He has a wicked smile that is not at all Disney-friendly, and then his face gradually falls apart piece by piece. It's obviously someone's idea of a joke that frankly isn't all that entertaining on either a psychological or humorous level. Still, if you want to investigate the story more, there's a surprising amount of information out there about it.

Every parent warns their children of the dangers of amusement park rides. These warnings begin the instant their kid can pass the "You must be this tall to ride..." line. The hope is that by instilling respect and a healthy dose of fear in their child, they'll avoid any issues while on vacation at a Disney park.

Does a wives' tale that provides a doomsday scenario for that philosophy help? That depends entirely on its extremity. In the case of Space Mountain, the rumor is so scary that it may frighten your child too much to relay it. That's your call.

According to legend, a misbehaving boy managed to slide his legs out from under the safety harness, allowing him to stand up. As you know, the ride primarily takes place in the dark, so the kid couldn't see the impending danger and, well, lost his head.

The story isn't true, of course. Space Mountain doesn't have that sort of danger intrinsic in its design. A person would have to work hard to find a way to get decapitated during the ride. That doesn't mean it's impossible, though. Silver Dollar City in Branson, Missouri, experienced this kind of incident in 1980. So, it's a lie with the ring of truth. Having said that, coasters move at such high velocity that collisions can cause loss of limb, as occurred at Alton Towers in the summer of 2015. As such, this myth works as a cautionary tale for misbehaving children, assuming they don't suffer nightmares after hearing the story.

An unexpected number of myths involve It's a Small World. I suspect it's because some folks despise the song so much that it elevates their desire to perform character assassination on the entire attraction as well as Disney as a company. Whatever the explanation, the Small World stories strain credulity.

The first of them falls in the haunted spirits category. A woman who took a boat ride in 1999 came away with a disturbing memory and an even more chilling photograph. She claims that her ride suddenly stopped when the entirety of It's a Small World shut down. It was the type of breakdown that required cast members to turn on the lights and tell guests that they were going to be extracted from the ride immediately.

One of the occupants being rushed out of the building grabbed her camera, sensing something unusual occurring. Remember that this was the age before the ubiquity of smart-phone photography. People had to wait weeks to develop film (Google it, kids), and when she finally got her vacation photos back, one of them was shocking. She claims that a child had hung itself in the rafters and she had unintentionally recorded it for posterity's sake.

It's not even a little bit true. The ride is virtually nothing but dolls that are child-like in appearance. There was no report of a death at any Disney theme park during this timeframe. And the blurry picture shot is a strong hint that somebody snapped a horrid picture that they'd now instantly delete off their phone. Since this was 1999, however, it became a scandalous turn of events that proved that a child haunted the rafters of It's a Small World.

Please don't shoot the messenger for this nonsense. There are people who believe that It's a Small World's dolls come to life at night. Imagine Toy Story as an amusement park attraction, and you get the gist. If you're anything like me, the mere thought of that will haunt your dreams for months.

All of those creepy dolls don't need power to operate. After cast members leave at night, the characters re-animate and throw a party, presumably talking about all the annoying humans who rode through that day. Perhaps Walt Disney's ghost or the immortal Walt Disney himself even visits to hang out with them. Is it okay if we blend multiple myths together at once like that?

Believe it or not, the story can and does get weirder. Some people aren't content with the concept of living dolls who act as statues during the day before moving around at night. They had to plus the urban legend even further. Their belief is that the cast members who loved It's a Small World return there when they die. Yes, the creepy dolls are possessed by the spirits of former Disney cast members, and they are the ones who enjoy the scary doll dance after park closing. Let's all agree never to ride or think about It's a Small World ever again.

Quirks regarding attractions, spooks who refuse to leave, and a frozen (or undead?) park founder are all fascinating myths, but there's a final category of nonsense so ridiculous that it boggles the mind. Here are the weird urban legends about Disney that are impossible to categorize.

Some people are deluded enough that they think Cinderella Castle at Magic Kingdom hides a secret. The state of Florida suffers from extreme weather, especially during hurricane season. Disney conspiracy theorists are convinced that Imagineers planned for the most catastrophic of conditions. If a hurricane ever came inland, Disney employees could unpack Cinderella Castle in hundreds of tidy pieces and relocate it to safety until the storm passed.

I...you...what? How does anybody even debunk that? Isn't common sense good enough as a refutation? There are castles that have survived for centuries that aren't as well fortified as Cinderella Castle. If *Red Dawn* ever became a real thing, Cinderella Castle would be a great place to stay until the enemy was pushed off American soil. It's that solid in its construction. A place that can survive cannon fire isn't going to disassemble like a Mr. Potato Head. What's there is there. Permanently. If anything, the hurricane would be a strong test of the 600-ton steel infrastructure on which the castle walls stand.

One of the joys of Disney theme parks is the training provided to certain cast members. The people provided the privilege of representing Disney's iconic characters in costume must remain as true as possible to the history and behavior of those characters. Mulan will act like a warrior, Dug will meet and instantly love you, and Mary Poppins will speak in an authentic accent. It's this dedication to detail that differentiates Disney's theme parks from basic amusement parks.

Sometimes, having to act in character is especially tricky. Take, for example, the cast of *Toy Story*. Like the It's a Small World dolls above, Woody and Buzz and the rest of their team only come to life when the humans aren't around. The recurring joke in the movie franchise is that the instant someone shouts a warning that their owner, Andy, is around, they must freeze in place.

Take that premise and extrapolate it to a theme park setting. Say, for example, that you see the army soldiers from *Toy Story* marching in formation toward a greeting site. What would happen if you shouted, "Everybody freeze! Andy's coming!" The only acceptable answer here is chaos.

Disney theme parks are precisely run operations requiring cast members to maintain order in the face of the unimaginable chaos of thousands of children under 10. You can't have people in giant costumes dropping to the ground in order to stay in character. It's a danger to everyone involved, and it creates mayhem for people in the surrounding area.

In order to avoid that eventuality, Disney has a standing directive that cast members cannot be 100 percent authentic when it comes to Toy Story quotes. If you shout, "Andy's coming," the mascot will acknowledge you. They will not, however, dive to the ground. At least, that's true today.

In 2013, Dewayne Bevil of the *Orlando Sentinel* tested the practice. He'd seen a viral image making the rounds that showed Woody and Jesse sitting lifeless after a guest had pointed out that Andy was on his way. A third person, presumably the person providing the warning, rests between them in order to get in on the fun.

Bevil, a highly respected, long-time amusement park reporter, received a reply from a Disney spokesperson. They acknowledged that due to safety concerns, Toy Story characters are no longer allowed to participate in what is a wonderful tribute to a basic theme of the movie franchise. The same cast member, however, refused to deny that it had never happened previously. That's the company's way of confirming that it seemed like an entertaining practice that proved a bit sketchy in practice, so Disney had to shut it down.

So, this myth is busted but also confirmed as historically accurate. If you ever manage to persuade a Disney cast member to perform this rite on a slow day, you are officially my hero. Because it's awesome.

Centuries before Stephen King and *The Simpsons Movie* celebrated the idea of living under a dome, this rumor existed regarding Epcot. Some people were convinced that Walt Disney picked Florida for a reason. While he couldn't control hurricane season, presumably due to a disappointing series of trials from Imagineering's Weather Division, Walt could make other plans.

The most legendary of these is that he intended to build a dome. Within its domain, Disney could guarantee perfect weather at all times thanks to the magic of what I'm sure would have been the world's largest central air unit. Seriously, that might have been the original purpose of the Spaceship Earth facility.

The belief that Disney intended to do this was so pervasive that it was as tethered to the concept of Epcot as the monorail and the PeopleMover. Any passion for this idea never made it out of the planning stage. While Imagineers did consider it alongside countless other potential innovations for the City of Tomorrow, they discarded the idea for the obvious reason. Do you have any idea how much the materials would cost for a dome of that size? And you're not going to find anything like this prefabricated at Home Depot. This explains why Orlando residents do not currently reside under the dome.

Have you heard the one about the Great Movie Ride? It's the tall tale that claims that in order to maximize the authenticity of the attraction's celebration of the glamor days of Hollywood, Disney imported a special artifact. That movie treasure is none other than the plane from *Casablanca*, the one that Ilsa boards for her flight to Lisbon, the one that permanently separates her from the love of her life, Rick.

While I admire the people who started this rumor for having tremendous movie taste, they didn't show a lot of filmmaking knowledge by starting it. Anyone who has ever worked on a set understands that producers have to cut a lot of corners to stay on budget. Few things are ever built to scale. In the case of *Casablanca*, one of the most storied films ever made, plane models were crafted on set, but they were not full-sized.

Someone at Disney missed this point, which led to a public relations blunder. See, the reason why this myth has been propagated so much is that it came directly from Disney. Almost exactly a decade after the Great Movie Ride debuted, the company wanted to drum up attention for one of their anchor attractions at Hollywood Studios. Employees performed an international search to unearth a plane similar enough to the one from *Casablanca* to pass for it. Whether they planned the misinformation campaign from the start or it happened accidentally is up for debate.

What we do know is this. The plane on display in the Great Movie Ride is a Lockheed Electra 12A. Amusingly, the one Disney managed to buy had been used on some film sets, just not for *Casablanca*. It was the right plane for the project, and had the company just stopped there, everyone would have praised them for their commitment to authenticity.

But Disney was in for a penny, in for a pound. They boldly proclaimed their coup. They had acquired the plane Humphrey Bogart stood beside when he sent Ilsa away, telling her that she'd regret it if she stayed, for the rest of her life. It's a magical movie moment, but Disney's part of the story was a lie. Their plane had nothing to do with the actual production of *Casablanca*. In fact, their Lockheed was a better quality vehicle than the Travel-Air used in the film. It was a rare black eye for the company, and their public announcement is the unlikely source of the myth. See? Even Disney can get things wrong when it comes to Disney history.

Acknowledgments

The content you've read in this book is collated from David Mumpower's writings at Theme Park Tourist. David previously extended his tremendous gratitude to Nick Sim for affording the author this opportunity. In a tragic turn of events, Nick died only a couple of days after the publication of *Disney Demystified: Volume One*. The world is a worse place in his absence, and the author misses his presence.

If you're unfamiliar with Nick's work, he was a gifted content creator who has two books available for purchase. You should read Universal Orlando: The Official Story and Tales from the Towers: The Unofficial Story Behind Alton Towers, Britain's Most Popular Theme Park when you have a chance.

David also thanks Amanda Kondolojy for her insights and feedback. Theme Park Tourist is unquestionably the best long-form theme park discussion site on the internet today, and it should become a part of the reader's daily internet surfing. All of Nick's work is saved for posterity on the site, which is now run by his wonderful wife, Natalie Sim.

The author also wants to thank his mother who—and this is not a joke—reads Theme Park Tourist so that she can correct any mistakes in her son's writing. His mother's passion is the reason why he's such a perfectionist in life.

This book is a loving tribute to David's father, who died after a 14-year battle with cancer in 2014. A lifelong employee of Eastman/Kodak, Howell Everette

Mumpower always admired the infrastructure that supported industry, and that's why he lovingly introduced his children to EPCOT Center soon after its debut. Had he chosen a different career path after he left the Navy, H.E. unquestionably would have become one of the finest writers in the world. These writings are a son's way of carrying on the legacy of the far-better man who raised him.

About the Author

David Mumpower has enjoyed his status as a premiere online content creator since 1998. He's written and published more than 10 million words in that time. Considered one of the finest movie analysts in the world, his work has been cited in publications such as CNN Money, Slate, Salon, Hitfix, io9, and *USA Today*.

David is the co-founder of the popular pop culture websites BoxOfficeProphets.com and HowWellDoYouKnow.com. He is blessed to work with some of the finest minds in the world through these two sites. To an individual, they're not just brilliant but also at the top of Maslow's Hierarchy. They are winning at life. David's only half a human without the love and support of all these incredible individuals.

David first visited Walt Disney World back when Future World was new, and his love of the original vision for the Experimental Prototype Community of Tomorrow remains to this day. His favorite ride is Spaceship Earth, a marvelous combination of architectural triumph and a historical re-creation of the dawn of man through modern times.

Today, David uses virtually all of his vacation days in Orlando, Florida, where he winds up spending all of his disposable income buying new Stitch merchandise for his wife. When he's in Orlando, you'll find him meticulously checking his Fitbit to figure out how many miles he's walked that day at the parks (14.6 is his current record). Obviously, he's a Park Hopper... with sore feet.

David's work is also available for consumption at DVC News and DVC Resale Market, where he enjoys strong working relationships with Tim Krasniewski and Nick Cotton. The extended Mumpower family is an ardent supporter of the Disney Vacation Club, which has allowed them to share several family vacations over the years.

About Theme Park Press

Theme Park Press publishes books primarily about the Disney company, its history, culture, films, animation, and theme parks, as well as theme parks in general.

Our authors include noted historians, animators, Imagineers, and experts in the theme park industry.

We also publish many books by first-time authors, with topics ranging from fiction to theme park guides.

And we're always looking for new talent. If you'd like to write for us, or if you're interested in the many other titles in our catalog, please visit:

www.ThemeParkPress.com

....

Theme Park Press Newsletter

Subscribe to our free email newsletter and enjoy:

- ♦ Free book downloads and giveaways
- ♦ Access to excerpts from our many books
- ♦ Announcements of forthcoming releases
- ♦ Exclusive additional content and chapters
- ♦ And more good stuff available nowhere else

To subscribe, visit www.ThemeParkPress.com, or send email to newsletter@themeparkpress.com.

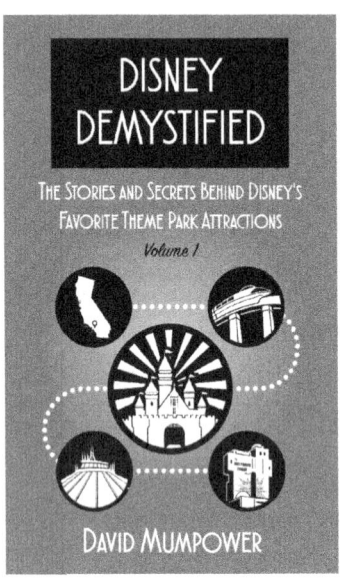

Read more about these books and our many other titles at:

www.ThemeParkPress.com

Made in the USA
Monee, IL
20 December 2019